SAS Manual

for

Moore and McCabe's
Introduction to the Practice
of Statistics

Third Edition

Michael Evans
University of Toronto

W. H. Freeman and Company
New York

Cover image: Andy Warhol, *200 Cambell's Soup Cans*, 1962. 72 × 100 inches.
© 1998 Andy Warhol Foundation for the Visual Arts/ASRS, New York, NY.
Photo courtesy of Leo Castelli Photo Archives.

ISBN 0-7167-3657-8

Printed in the United States of America

First printing 1999

Contents

Preface   ix

I   SAS Language   1

1   Overview and Conventions 3
2   Accessing and Exiting SAS 4
3   Getting Help 7
4   SAS Programs 9
5   Data Step 13
    5.1   Form and Behavior of a Data Step 14
    5.1   SAS Constants, Variables, and Expressions 15
    5.3   Input 17
    5.4   Output 27
    5.5   Control Statements 32
6   SAS Procedures 37
    6.1   PROC PRINT 37
    6.2   PROC SORT 38
7   Exercises 41

II   SAS Procedures for Data Analysis   43

1   Looking at Data: Distributions   45
    1.1   Tabulating and Summarizing Data 46
      1.1.1   PROC FREQ 46
      1.1.2   Calculating the Empirical Distribution Function 49
      1.1.3   PROC MEANS 50

    1.1.4   PROC UNIVARIATE   56
1.2 Graphing Data   60
    1.2.1   PROC CHART   60
    1.2.2   PROC TIMEPLOT   65
1.3 Graphing Using SAS/Graph   66
    1.3.1   PROC GCHART   67
1.4 Normal Distribution   67
    1.4.1   Normal Quantile Plots   69
1.5 Exercises   70

2 Looking at Data: Relationships   **73**
2.1 Relationships Between Two Quantitative Variables   73
    2.1.1   PROC PLOT and PROC GPLOT   73
    2.1.2   PROC CORR   78
    2.1.3   PROC REG   83
2.2 Relationships Between Two Categorical Variables   87
2.3 Relationship Between a Categorical
     Variable and a Quantitative Variable   90
    2.3.1   PROC TABULATE   91
2.4 Exercises   94

3 Producing Data   **97**
3.1 PROC PLAN   98
3.2 Sampling from Distributions   99
3.3 Simulating Sampling Distributions 101
3.4 Exercises 102

4 Probability: The Study of Randomness   **105**
4.1 Basic Probability Calculations 105
4.2 Simulation 107
    4.2.1   Simulation for Approximating Probabilities 107
    4.2.2   Simulation for Approximating Means 108
4.3 Exercises 109

5 From Probability to Inference   **113**
5.1 Binomial Distribution 113
5.2 Control Charts 115
    5.2.1   PROC SHEWHART 115

5.3   Exercises 119

6 Introduction to Inference   123
6.1 z Intervals and z tests 123
6.2   Simulations for Confidence Intervals 126
6.3   Simulations for Power Calculations 128
6.4   Chi-square Distribution 131
6.5   Exercises 133

7 Inference for Distributions   135
7.1   Student Distribution 135
7.2   The t Interval and t Test 136
7.3   The Sign Test 139
7.4   PROC TTEST 140
7.5 F Distribution 143
7.6   Exercises 144

8 Inference for Proportions   147
8.1   Inference for a Single Proportion 147
8.2   Inference for Two Proportions 149
8.3   Exercises 152

9 Inference for Two-way Tables   153
9.1   PROC FREQ with Nontabulated Data 153
9.2   PROC FREQ with Tabulated Data 156
9.3   Exercises 157

10 Inference for Regression   159
10.1 PROC REG   159
10.2 Example 161
10.3 Exercises 165

11 Multiple Regression   167
11.1 Example Using PROC REG 167
11.2 PROC GLM 170
11.3 Exercises 172

12 One-way Analysis of Variance   **173**
   12.1 Example 174
   12.2 Exercises 178

13 Two-way Analysis of Variance   **181**
   13.1 Example 181
   13.2 Exercises 187

14 Nonparametric Tests   **189**
   14.1 PROC NPAR1WAY 189
   14.2 Wilcoxon Rank Sum Test 190
   14.3 Sign Test and Wilcoxon Signed Rank Test 192
   14.4 Kruskal-Wallis Test 193
   14.5 Exercises 194

15 Logistic Regression   **197**
   15.1 Logistic Regression Model 197
   15.2 Example 198
   15.3 Exercises 202

III   Appendices   **205**

A Operators and Functions in the Data Step   **207**
   A.1 Operators 207
      A.1.1 Arithmetic Operators 207
      A.1.2 Comparison Operators 208
      A.1.3 Logical Operators 208
      A.1.4 Other Operators 208
      A.1.5 Priority of Operators 208
   A.2 Functions 209
      A.2.1 Arithmetic Functions 209
      A.2.2 Truncation Functions 209
      A.2.3 Special Functions 210
      A.2.4 Trigonometric Functions 210
      A.2.5 Probability Functions 210
      A.2.6 Sample Statistical Functions 211
      A.2.7 Random Number Functions 212

B Arrays in the Data Step **213**

C PROC IML **217**
    C.1 Creating Matrices 217
        C.1.1 Specifying Matrices Directly in IML Programs 217
        C.1.2 Creating Matrices from SAS Data Sets 218
        C.1.3 Creating Matrices from Text Files 220
    C.2 Outputting Matrices 220
        C.2.1 Printing 220
        C.2.2 Creating SAS Data Sets from Matrices 221
        C.2.3 Writing Matrices to Text Files 221
    C.3 Matrix Operations 222
        C.3.1 Operators 222
        C.3.2 Matrix-generating Functions 228
        C.3.3 Matrix Inquiry Functions 229
        C.3.4 Summary Functions 230
        C.3.5 Matrix Arithmetic Functions 231
        C.3.6 Matrix Reshaping Functions 232
        C.3.7 Linear Algebra and Statistical Functions 233
    C.4 Call Statements 236
    C.5 Control Statements 238
    C.6 Modules 242
    C.7 Simulations Within IML 244

D Advanced Statistical Methods in SAS **245**

E References **247**

Index **249**

Preface

This SAS manual is to be used with *Introduction to the Practice of Statistics*, Third Edition, by David S. Moore and George P. McCabe, and to the CD-ROM that accompanies this text. We abbreviate the textbook title as IPS.

SAS is a sophisticated computer package containing many components. The capabilities of the entire package extend far beyond the needs of an introductory statistics course. In this book we present an introduction to SAS that provides you with the skills necessary to do all the statistical analyses asked for in IPS and also sufficient background to use SAS to do many of the analyses you might encounter throughout your undergraduate career. While the manual's primary goal is to teach SAS, more generally we want to help develop strong data analytic skills in conjunction with the text and the CD-ROM.

The manual is divided into three parts. Part I is an introduction that provides the necessary details to start using SAS and in particular discusses how to construct SAS programs. The material in this section is based on references 1 and 2 in Appendix E. Not all the material in Part I needs to be fully absorbed on first reading. Overall, Part I serves as a reference for many of the nonstatistical commands in SAS.

Part II follows the structure of the textbook. Each chapter is titled and numbered as in IPS. The last two chapters are not in IPS but correspond to optional material included on the CD-ROM. The SAS procedures (proc's) relevant to doing the problems in each IPS chapter are introduced and their use illustrated. Each chapter concludes with a set of exercises, some of which are modifications of or related to problems in IPS and many of which are new and specifically designed to ensure that the relevant SAS material has been understood. The material in this part is based on references 3, 4, 5 and 6 in Appendix E.

We recommend that you read Part I before starting Chapter 1 of Part II. Sections I.5.3, I.5.4, and I.5.5 do not need to be read in great detail the first time through. Part I together with the early chapters of Part II represent a fairly heavy investment in study time but there is a payoff, as subsequent chapters are much easier to absorb and less time is required. Again, each chapter in Part II contains more material than is really necessary to do the problems in IPS. In part, they are to serve as references for the various procedures discussed. So the idea is not to read a chapter with the aim of absorbing and committing to memory every detail recorded there, but rather get a general idea of the contents and the capabilities of each procedure. Of course you want to acquire enough knowledge to do the problems in IPS using SAS. It is recommended that you use SAS to do as many of the problems as possible. This will ensure that you become a proficient user of SAS.

Part III contains Appendices dealing with more advanced features of SAS, such as matrix algebra. Appendices A and B are based on more advanced material from references 1 and 2 in Appendix E. Appendix C is based on reference 7 in Appendix E. This material may prove useful after taking the course, so it is a handy reference and hopefully an easy place to learn the material. Appendix D lists some of the more advanced statistical procedures found in references 3 and 4.

SAS is available in a variety of versions and for different types of computing systems. In writing the manual we have used Release 6.12 for Windows. For the most part the manual should be compatible with other versions of SAS as well.

Many thanks to Patrick Farace, Chris Granville, and Chris Spavens of W. H. Freeman for their help. Also thanks to Rosemary and Heather for their support.

Part I
SAS Language

SAS statements introduced in this part

by	file	libname	proc sort	storage
cards	goto	list	put	update
contents	id	lostcard	rename	var
data	if-then-else	merge	run	
datalines	infile	missing	select-otherwise	
do-end	input	output	set	
drop	keep	proc print	stop	

1 Overview and Conventions

Part I is concerned with getting data into and out of SAS and giving you the tools necessary to perform various elementary operations on the data so that they are in a form in which you can carry out a statistical analysis. You don't need to understand everything in Part I to begin doing the problems in your course. But before you start on Part II, you should read Part I, perhaps reading sections 5.3-5.5 lightly and returning to them later for reference.

SAS is a software package that runs on several different types of computers and comes in a number of versions. This manual does not try to describe all the possible implementations or the full extent of the package. We limit our discussion to aspects of version 6.12 of SAS running under Windows 95/98/NT. Also, for the most part, we present only those aspects of SAS relevant to carrying out the statistical analyses discussed in IPS. Of course, this is a fairly wide range of analyses, but the full power of SAS is nevertheless unnecessary. The material in this manual should be enough to successfully carry out, in any version of SAS, the analyses required in your course.

In this manual special statistical or SAS concepts are highlighted in *italic* font. You should be sure you understand these concepts. We provide a brief explanation for any terms not defined in IPS. When a reference is made to a SAS *statement* or *procedure* its name is in **bold** face. *Menu commands* are accessed by clicking the left button of the mouse on items in lists. We use a special notation for menu commands. For example,

A ▶ B ▶ C

means left-click the command A on the menu bar, then in the list that drops down left-click the command B and finally left-click C. The menu commands are denoted in ordinary font exactly as they appear. The record of a SAS session — the commands we type and the output obtained — are denoted in `typewriter` font, as are the names of any files used by SAS, variables, and constants.

SAS is case insensitive except for the values of character variables. The statement `PROC PRINT;` has the same effect as `proc print;` and the variable `a` is the same as the variable `A`. However, if `a` and `b` are character variables that take the values `X` and `x` for a particular observation, then they are not equal. For convenience we use lower case whenever possible except, of course, when referring to the values of character variables that include upper case.

At the end of Part I and of each chapter in Part II we provide a few exercises that can be used to make sure you have understood the material. We also recommend that whenever possible you use SAS to do the problems in IPS. While many problems can be done by hand you will save a considerable amount of time and avoid errors by learning to use SAS effectively. We also recommend that you try out the SAS concepts and commands as you read about them to ensure full understanding.

2 Accessing and Exiting SAS

The first thing you should do is find out how to access the SAS package for your course. This information will come from your instructor or system personnel, or from software documentation if you have purchased SAS to run on your own computer. If you are not running SAS on your own system, you will probably need a login name and a password to the computer system being used in your course. After you have logged on — provided a login and

SAS

Figure 1: Short-cut icon to SAS software.

password to the computer system — you then access SAS.

Under Windows 95/98/NT you can access SAS in a number of ways. Perhaps the easiest method is to left-click or double left-click, whichever is relevant to your system, an icon such as the short-cut icon in Figure 1.

Alternatively you may use the Start button on the task bar in the lower left-hand corner of your screen: Start ▶ Run, fill in the dialog box that pops up with

```
C:\SAS\sas.exe
```

and hit Enter. This gives the pathname for the SAS program, which here resides on the C drive in the folder called SAS. If the software resides in a different folder, then a different pathname must be given. In either case, once you have invoked SAS, the *Display Manager window* shown in Figure 2 should be displayed on your screen.

The top line of the Display Manager window is called the *menu bar*. It contains File, Edit, View, Locals, Globals, Options, Window, and Help when the Program Editor window is active and File, Edit, View, Globals, Options, Window and Help when the Log window is active. A window is made active by clicking on it. The upper border of an active window is dark blue; that of an inactive window is gray. Left-clicking any of the commands in the menu bar produces a drop-down list. Below the menu bar is a small window where text can be typed and a row of buttons called the *task bar*. The buttons in the task bar correspond to frequently used commands that can also be accessed from the drop-down lists in the menu bar.

Below the task bar is the *Log window*. When you run a SAS program, a listing of the program is printed in this window. Details concerning the running of the program are printed here, e.g. how much time it took. If the program doesn't work, SAS prints messages along with the program, that indicate where it found errors. The Log window should always be examined after running a SAS program.

The *Program Editor window* appears below the Log window. It is here

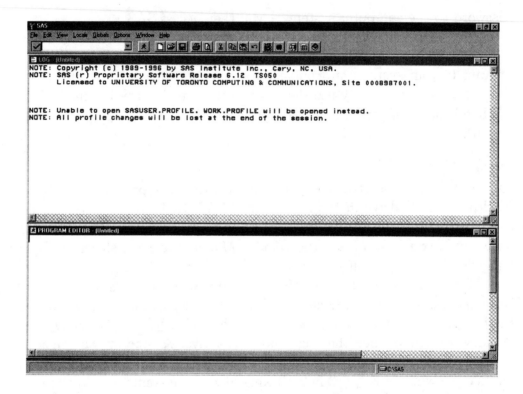

Figure 2: SAS Display Manager window containing menu bar, task bar, Log window and Program Editor window.

that the various commands that make up a SAS program are typed. Once a SAS program is typed into the Program Editor window it is submitted for execution. We describe action this in Section I.3.

If the program runs successfully then the output from the program will appear in the *Output window*. To access the Output window use the menu command Window ▶ 3 OUTPUT. Notice that you can toggle between windows by using Window ▶ 1 LOG, Window ▶ 2 PROGRAM EDITOR and Window ▶ 3 OUTPUT. If you use the command Window ▶ Cascade then all three windows are displayed in the Display Manager window, overlaying one another, with the active one on top. You can then toggle between the windows by left-clicking in the window you wish to be active. Clicking anywhere on the upper border of a window will maximize the window to fill the entire Display Manager window. Note that you can also toggle between the windows by clicking on the relevant item in the drop-down list that is produced from Globals ▶ .

To exit SAS you can simply click on the close window symbol in the upper right-hand corner of the Display Manager window. You are then asked if you really want to terminate your SAS session. If you respond by clicking OK, then the Display Manager window closes and the SAS session ends. Of course, any programs, logs, or output produced during the SAS session are lost unless they have been saved. We subsequently discuss how to save such data so that it can be reused. Alternatively you can use the menu command File ▶ Exit with the same results.

3 Getting Help

At times you may want more information about a command or some other aspect of SAS than this manual provides, or you may wish to remind yourself of some detail you have partially forgotten. SAS contains a convenient online help manual. There are several ways to access help. The simplest method is to click on the help button in the task bar depicted in Figure 3.

Figure 3: The Help button.

Alternatively use the menu command Help ▶ Online documentation. Ei-

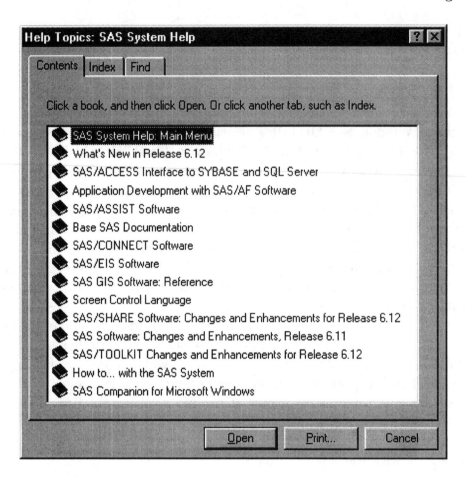

Figure 4: Help window.

ther of these opens a help window, as depicted in Figure 4. Observe that there are three tabs. The first tab is called Contents and gives a list of major topics in the online manual, each with a book icon beside it. Click on any of these and then click on Open at the bottom of the Help window. This causes a further list of topics to be displayed. Documents describing features of SAS are marked with the document icon (Figure 5); a book icon leads to a further set of topics. Clicking on a document icon and the Display button at the bottom of the Help window causes the text of the document topic to be displayed.

This method is fine when we want to read about large topics. Often, however, we want to go directly to the point in the manual where a specific

Figure 5: Document icon.

topic is discussed, for example, the **input** statement. For this we click on the Index tab and type "input" into the small window that appears there. The section of the manual that discusses this statement immediately opens. If there is more than one section relevant to what we typed, the window scrolls to a place in the list where these entries are listed. Clicking on one of them and then on the Display button causes the text to be displayed.

If the topic we are interested in cannot be easily located in the Contents or Index, then we can use the Find tab to search the manual for entries relevant to the word or phrase of interest. This is often the most convenient way to call up information about a topic.

Conveniently there are hyperlinks throughout the manual that allow you to navigate your way through the manual via related topics. Spend some time getting acquainted with Help. Undoubtedly you will need to use it at some time.

4 SAS Programs

A SAS program typed into the Program Editor window is submitted for processing by clicking on the Submit button shown in Figure 6. Alternatively you can use the menu command Locals ▶ Submit.

Figure 6: Submit program button.

A SAS program consists of *SAS statements* and sometimes data. Each statement must end in a semicolon. A SAS statement can be placed on more than one line. If a statement does not scan correctly as a valid SAS statement, then an error will occur and the program will not run. Broadly speaking the statements in a SAS program are organized in groups in two categories: *data steps* and *procedures* (hereafter referred to as proc's, as is standard when discussing SAS). Essentially the data steps are concerned with constructing

SAS data sets that are then analyzed via various procedures. The data step typically involves inputting some data from a source, such as an external file, and then manipulating the data so that they are in a form suitable for analysis by a procedure. In an application we will have identified some SAS statistical procedures as providing answers to questions we have about the real-world phenomenon that gave rise to the data. There are many different SAS procedures, and they carry out a wide variety of statistical analyses. After a SAS procedure has analyzed the data, output is created. This may take the form of (a) actual output written in the Output window (b) data written to an external file, or (c) a temporary file holding a SAS data set.

Let us look at a very simple SAS program. Suppose we type the following commands in the Program Editor window:

```
data;
input x;
cards;
1
2
3
proc print;
var x;
run;
```

The data step consists of all the SAS statements starting with the line `data;` and ending with the line `cards;`. The **cards** statement tells SAS that this data step is over. The word **cards** is a holdover from the days when programs were submitted to computers on punch cards. Alternatively we can use the word **datalines** instead of **cards**. Immediately following the end of the data step are the data; in this case there are three observations, or *cases*, where each observation consists of one variable x. The value of x for the first observation is 1, for the second 2, and for the third 3. SAS constructs this data set and calls it `work.data1`. This is a default name until we learn how to name data sets.

Immediately after the data comes the first procedure, in this case the **proc print** procedure. SAS knows that the actual data have ended when it sees a word like **data** or **proc** on a line. The procedure **proc print** consists of the line `proc print;` and the subsequent line `var x;`. The **var** statement tells **proc print** which variables to print in the just created SAS data set `work.data`. Since there is only one variable in this data set, there is really no need to include this statement. As we will see, we can also tell a proc which SAS data set to operate on; if we don't, the proc will operate on the most recently created data set by default.

The **run** command tells SAS that everything above this line is to be executed. If no **run** command is given, the program does not execute although the program is still checked for errors and a listing produced in the Log window. Prior to submission, the Program Editor window looks like Figure 7.

Figure 7: Program Editor window.

After submitting the program, the Program Editor window empties and a listing of the program together with comments is printed in the Log window displayed in Figure 8. We see from the Log window that the program ran, so we then check the Output window for the output from the program (Figure 9). Notice that the Output window contains the three observations.

Suppose you submit your program and you have made an error. Because the Program Editor window empties after submission, you have to put the originally submitted program back into this window and correct it. This is done by the menu command Locals ▶ Recall text when the Program Editor window is active. If you forgot to type in the **run** command, then the listing and any error messages based on the syntax of SAS commands only, no error messages based on the execution of the program, are produced in the Log window. If a program has a long running time, this is a good way to scan the program for syntax errors before actually running it.

If you have text in any of the windows and you don't want it there, then the window can be emptied by activating the window and using the command Edit ▶ Clear text. Alternatively, you may want to print the contents of a window. This is done by making the relevant window active and then using the command File ▶ Print. If you want to save the contents of a window to a

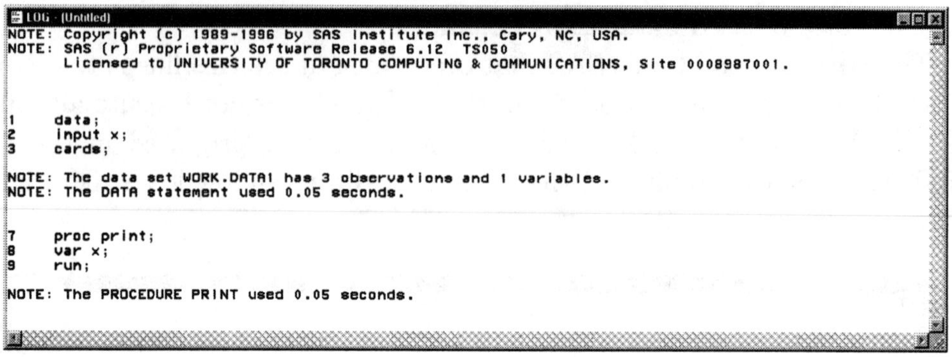

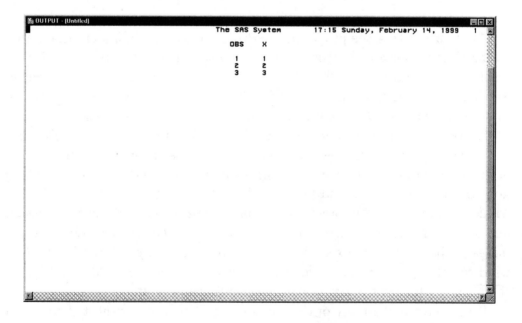

Figure 8: Log window.

Figure 9: Output window.

file, use the command File ▶ Save or File ▶ Save as. A window pops up and
you are asked which folder you want to save the window contents in, what
name you want to give the file and what suffix you want to give the file name.
For example, you can choose the suffices .sas, .log, .lst, .dat, or .rtf. If you
choose .dat or .rtf, then the contents of the window are placed in a text file,
and clicking on the file is likely to result in some editor on the system opening
the file to be edited. The .sas ending, however, is treated in a special way, as
this suffix identifies the contents of the file as consisting of a SAS program
(whether it is a valid program or not). Clicking on such a file causes the SAS
program to launch and the contents of the file to be loaded into the Program
Editor window. Note that this gives you an alternative way of constructing
SAS programs: use your favorite editor and then save the SAS statements
in a file with the suffix .sas. If you use the .lst suffix, clicking on the file
results in it being opened by the *SAS System Viewer*, provided this software
has been installed on your system. The System Viewer regards .lst files as
output from a SAS program that has been written to the Output window.
If you use the .log suffix, clicking on the file results in it being opened by
the System Viewer, which regards the file as a listing of a program, together
with comments, that has been written to the Log window.

A SAS program can also contain *comments.* This is helpful when you have
a long program and you want to be able to remind yourself later about what
the program does and how it does it. Comments can be placed anywhere in
a SAS program provided they start with /* and end with */. For example,

```
data; /* this is the data step */
input x;
cards;
1 /* this is the data */
2
2
proc print; /* this is a procedure */
run;
```

5 Data Step

We discuss only the key features of the data step in this section. For
the many other more advanced procedures, we refer the reader to reference
1 in Appendix E. In Appendix B we discuss *arrays,* which are useful when

carrying out extensive data processing that requires recoding of data. This material is not needed in an elementary statistics course, however.

5.1 Form and Behavior of a Data Step

The first thing you have to know how to do is to get your data into the SAS program so that they can be analyzed. SAS uses data steps to construct *SAS data sets*. Several SAS data sets can be constructed in a single SAS program by including several data steps. These data sets can be combined to form new SAS data sets, and they can be permanently stored in computer files so that they can be accessed at a later date.

The general form of the data step is

data *name*;
statements
cards;

where *name* corresponds to a name we give to the SAS data set being constructed and *statements* is a set of SAS statements. These statements typically involve reading data from observations from some source and perhaps performing various mathematical operations on the data to form new variables. SAS expects to read data from some source whenever there is an **input** statement included in *statements*. For example, the default method of supplying the data to an **input** statement is to include the data in the program immediately after the **cards** statement. If we are not going to use this method, and it is often inconvenient because it may involve a lot of error-prone typing, then we must tell SAS where to find the data. We don't have to supply *name,* but if we don't, a default name is assigned, with the first unnamed SAS data set in the program called `work.data1`, the second called `work.data2`, and so on. In general it is better to name a data set with some evocative name so that you can remember what kind of data it contains. The value given to *name* must conform to certain rules, it must be a valid *SAS name*. A SAS name must begin with a letter or undescore _, can consist of letters, numbers, and the underscore _, and can be no longer than eight characters. For example, `x1`, `ab_c`, and `lemon` are all valid SAS names.

If there is no **input** statement, then the *statements* are executed and the data set *name* consists of one observation containing any variables introduced in *statements*. If an **input** statement is given in *statements,* then SAS uses an indexing variable _N_ in the following fashion. Initially it is given the value

_N_=1, the first observation is read from the source, and each statements in *statements* is executed. Then the assignment _N_=_N_+1 is made, the next observation is read, and each statement in *statements* is again executed. This continues until the last observation has been read from the data source. So you can see that a data step behaves like an implicit loop, and it is important to remember this. For example,

```
data one;
input x y;
z=x+y;
cards;
1 10
3 11
proc print;
run;
```

construct a SAS data set with two observations and three variables x, y, and z where x and y are read in from the data supplied in the program and the new variable z is constructed for each observation by adding its x and y values. The **print** procedure outputs

```
OBS X  Y  Z
 1  1 10 11
 2  3 11 14
```

in the Output window. The automatic variable _N_ can be referenced in the data step if we want to change the behavior of the data set for certain observations.

5.2 SAS Constants, Variables, and Expressions

There are two types of *SAS constants*. A *numeric constant* is simply a number that appears in a SAS statement. Numeric constants can use a decimal point, a minus sign, and scientific notation. For example, $1, 1.23, 01, -5, 1.2E23, 0.5E-10$ are all valid constants that can appear in a SAS program. A *character constant* consists of 1 to 200 characters enclosed in single quotes. For example,

```
data;
name='tom';
cards;
```

```
proc print;
run;
```

creates a SAS data set with a single observation with one variable called `name` that takes the value `tom`.

SAS variables are given SAS names, as described in Section 5.1. When several variable names all have the same beginning but differ by a final number that increases by 1, such as in `x1, x2, ... , x100`, then the entire list may be referenced as a group, called a *range list*, as in `x1-x100`. This can be a significant convenience, as we avoid having to type in long lists of variables when entering a SAS program.

SAS variables are either *numeric* or *character* in type. The type is determined by an **input** statement or an *assignment statement*. Numeric variables take real number values. For example, the assignment statement

```
x = 3;
```

assigns the value 3 to the numeric variable `x`. When a numeric variable exists in SAS but does not have a value, the value is said to be *missing*. SAS assigns a period as the value of the variable in this case. For example, the statement

```
x = .;
```

establishes that the variable `x` in the SAS data set being constructed is a numeric variable and is missing its value. It is very important in the analysis of data that you pay attention to observations with missing values. Typically, observations with missing values are ignored by SAS procedures.

Character variables take character values. For example, the assignment statement

```
x = 'abc';
```

establishes that `x` is a character variable taking the value `abc`. If a character variable has no value (is missing), then SAS assigns a blank to the value, for example,

```
x = ' ';
```

establishes that the variable `x` in the SAS data set being constructed is a character variable missing its value. In general we try to avoid the use of character variables, as handling them is troublesome, but sometimes we need to use them. For example, if you have a variable in a data set to denote an individual's gender, ,use 0 and 1 to distinguish between male and female rather than `male` and `female` or `M` and `F`.

A *SAS expression* is a sequence of constants, variables, and operators that determines a value. For example, in the statement

    x=y+z;

the numeric variable x is formed by adding the numeric variables y and z in the *arithmetic expression* y+z. The expression

    x<y;

takes the value 1 when the variable x is less than y and the value 0 otherwise; i.e., it takes the value 1 when the *logical expression* x<y is true and the value 0 otherwise.

A variety of operators can be used in SAS expressions. There are arithmetic operators such as addition +, subtraction −, multiplication *, division / and exponentiation **. There are comparison operators such as <, >, <=, >= and logical operators such as & (and), | (or) and ~ (not). For a full list of the operators available in SAS, see Appendix A.1. Also detailed there is the priority of the operators. If we use a SAS expression in a program, then we have to be concerned about how it is going to be evaluated; what is the value of $3 − 2/4, 2.5$ or $.25$ (answer: 2.5 because of the priority of the operators)? The simplest way to avoid having to remember the priority of the operators is to use parentheses (); e.g., write $3 − (2/4)$, as the expressions inside parentheses are always evaluated first. The value of an expression involving comparison operators or logical operators is 1 or 0 depending on whether the expression is true or false.

There is also a variety of a functions available in SAS, such as **sin**, **cos**, **log** (base e), **exp**. These are useful for forming new variables. For a complete listing see Appendix A.2. There are arithmetical functions such as sqrt(x) which calculates the nonnegative square root of x and special functions such as the Gamma function. Of some importance to this course are the probability functions that allow you to calculate cumulative distribution functions (cdf's) and inverse cumulative distribution functions for various distributions. For example,. probnorm(x) calculates the $N(0, 1)$ distribution function at x and probit(x) calculates the inverse distribution function for the $N(0, 1)$ distribution at x.

5.3 Input

We now describe more fully how to input data using the **input** statement. We identify three methods: from observations placed in the program, from external text files, and from other SAS data sets. In Subsections I.5.3.1, I.5.3.2, and I.5.3.4 we describe these methods of input using a restricted form of *list input*. By this we mean that each observation is on a single line and the values of its variables are separated by spaces. This is the simplest kind of input, but it is not always possible. In Subsection I.5.3.4 we describe more general forms of input.

5.3.1 Data Input from the Program

Suppose each observation consists of three variables, y = weight, x1 = height, and x2 = average number of calories consumed daily. Suppose we have four observations given by

```
160 66   400
152 70   500
180 72 4500
240 68 7000
```

The statements

```
data example;
input y x1 x2;
cards;
160 66   400
152 70   500
180 72 4500
240 68 7000
```

create a SAS data set named **example**, containing four observations, each having three variables y, x1, and x2.

5.3.2 Data Input from a Text File

Suppose we have previously created a text file called `C:\datafile.txt`, where we have provided its full pathname, and suppose it contains the data

```
160 66   400
152 70   500
180 72 4500
240 68 7000
```

To access this file for input we use the **infile** statement. The program

```
data example;
infile 'C:\datafile';
input y x1 x2;
cards;
```

reads these data into the SAS data set `example` and creates the SAS data set `example`.

5.3.3 Data Input from a SAS Data Set

Suppose we have a SAS program that creates a number of distinct data sets, and we want to combine the data sets to form a larger SAS data set. We do this using the **set** statement. For example, if the file `C:\stuff.txt` contains

```
1 2
3 4
```

then the SAS program

```
data one;
infile 'C:\stuff.txt';
input x y;
cards;
data two;
input x y;
cards;
5 6
data three;
set one two;
proc print data = three;
```

creates a new SAS data set called **three** that is the concatenation of **one** and **two**, i.e., the observations of **one** followed by the observations of **two**. The SAS data set **three** contains two variables **x** and **y** and three observations.

We can also use **set** to construct a data set consisting of a subset of the observations in a data set. The program

```
data one;
infile 'C:\stuff.txt';
input x y;
cards;
```

```
data two;
set one;
if y = 2;
```

creates a data set **two** from **one** by selecting only those observations for which
y = 2. Therefore **two** has only one observation. This is called a *subsetting if*
statement.

If data set **one** contains variables x1, x2, ..., z and data set **two** contains
variables y1, y2, ..., z and these data sets are sorted by z (see Section I.6.2),
then

```
data three;
set one two;
by z;
```

creates a new data set **three** that contains all the observations in **one** and
two, interleaved. By this, we mean that the observations in **three** occur in
the following order: all observations in **one** with the first value of z, then
all observations in **two** with the first value of z, then all observations in **one**
with the second value of z, and so on. When there are many observations
in **one** say with the same value of z, they are listed in **three** with the same
order they had in **one**. This is called *interleaving data sets*.

While **set** concatenates two data sets vertically, we can also concatenate
data sets horizontally using **merge**. The program

```
data one;
infile 'C:\stuff.txt';
input x y;
cards;
data two;
input z;
cards;
5
6
data three;
merge one two;
proc print data = three;
```

creates a data set **three** with two observations and three variables x, y,
and z by taking the horizontal union of **one** and **two**; then prints the result.
This is called *one-to-one merging*. Of course, the first observation in **one** is

matched with the first observation in **two**, so make sure that this makes sense in a particular application.

When using **set** or **merge**, it may be that you do not wish to keep all the variables. For example, suppose a SAS data set **one** contains w, x, y, and z. Then the statements

```
data two;
set one;
keep w x;
```

forms a new data set **two** from **one** that has the same observations but only two variables. The **keep** statement drops all variables but those named in the statement. The **drop** statement keeps all variables except those named in the statement. For example,

```
data two;
set one;
drop w;
```

creates data set **two** with the same observations as **one** but without the w variable values. The **drop** and **keep** statements can also appear as options in the **data** statement. For example,

```
data two (drop = x) three (keep = x);
set one;
```

creates two data sets from data set **one**. Data set **two** has variables w, y, and z and **three** has only variable x. The **keep** and **drop** statements help cut down the size of SAS data sets. This is important when we have only a limited amount of memory or have to pay for permanent storage.

It is also possible to rename variables in the newly created SAS data set using the **rename** statement. For example,

```
data two;
set one;
keep w x;
rename x = mnx;
```

creates a data set **two** with two variables w and **mnx**, where **mnx** takes the values of x.

While we have described the operation of the database commands **set** and **merge** on only two SAS data sets at a time, they can operate on any number.

For example, the statement `set one two three;` vertically concatenates data sets `one`, `two`, and `three`.

As we will describe in Subsection I.5.4, it is possible to write a SAS data set out to a permanent file. Such files end in the suffix .sd2, which identifies them as SAS data sets. These are formatted files and they cannot be edited except by using SAS tools. Suppose we have written a SAS data set to the file `C:\three.sd2`. Then the statements

```
libname storage 'C:\';
data one;
set storage.three;
cards;
proc print;
run;
```

read this data set into a SAS data set named `one` and print out its contents. Note that we did not need to include the suffix .sd2 in the `set` statement. Also we could have used any directory, other than the root directory on the C drive, as the **storage** directory. We have to specify the full pathname for this directory in the **libname** statement, and, of course, the file we wish to read has to be there. In this example we do not do anything with the data set read in. Alternatively, as we will see, we can operate directly on stored SAS data sets using proc's.

5.3.4 More General Input

In the examples considered so far, we have always used list input where each observation occupied only one row, data values of variables are separated by blanks, and missing values are represented by a period. In fact, we have only considered the input of numeric variables. While this simple form of input works quite often, it is clear that in some contexts more elaborate methods are required.

First we deal with character data. Recall that a character constant is a sequence of characters; e.g., `x = 'a15b'` specifies that `x` is a character variable and that it takes the value a15b. This is how we specify character variables in assignment statements. If we want to read in the values of a character variable in a data set, a $ sign must follow the name of the variable in the **input** statement: `input x $;`. We can then input character variables just as with numeric variables with blanks delimiting the values of the variable.

5.3.4.1 Pointer

As SAS reads an input record, it keeps track of its position, as it scans the record, with a *pointer*. Various commands can be used to change the position of the pointer. Here are some of them:

@n moves the pointer to column n
+n moves the pointer n columns
#n moves the pointer to line n

We discuss later how to use pointer commands to read in various types of input records.

5.3.4.2 List Input

With *list input,* variable names occur in the **input** statement in the order in which they appear in the data records. Values in data records are separated by blanks unless some other character is specified as a **delimiter** in an **infile** statement. For example, if the data is stored in a file C:\datafile.txt and the variable values are separated by commas, then the **infile** statement must be changed to

```
infile 'C:\datafile.txt' delimiter = ',';
```

Any other character can be used to separate data values. Missing values are represented by periods for all variables. Suppose the file C:\records.txt contains the data

```
A 1
3
BB 3
4
```

which we want to be read in as two observations with three variables x, y, and z, with x a character variable. The program

```
data example;
infile 'C:\records.txt';
input x $ y #2 z;
cards;
proc print;
```

does this and prints out the values; namely, in the first observation x = A, y = 1, z = 3, and in the second observation x = BB, y = 3, and z = 4. Notice that the pointer control # allows us to read in observations that occupy more than one line in the file.

5.3.4.3 Column Input

With *column input*, the values of variables occupy certain columns in the data record and not others. You must tell SAS what these columns are in the **input** statement. Blanks or periods alone are interpreted as missing values. Leading and trailing blanks around a data value are ignored. Suppose the file C:\datafile.txt contains two records containing two variables, where the first variable is in columns 1 to 10 and the second variable is in columns 15 to 20 and both are numeric. Suppose the actual file contains two lines, where the first line has 6.6 in columns 5, 6, and 7 and 234.5 in columns 16, 17, 18, 19, and 20 and the second line has 1345 in columns 2, 3, 4, and 5 and 678 in columns 11, 12, and 13. Then the program

```
data example;
infile 'datafile' pad;
input x 1-10 y 15-20;
cards;
```

reads in two observations where x = 6.6, y = 234.5 for the first observation and x = 1345, y = . (missing) for the second observation. Note that the **pad** option in **infile** is needed when the length of the values varies from record to record.

With column input, we can also specify the number of decimal places a value will have. For example,

```
input x 1-10 .3 y 15-20 .2;
```

with the previous data gives x = 6.600, y = 234.50 for the first observation and x = 1.345, y = . for the second.

5.3.4.4 Formatted Input

Sometimes list and column input are not possible or convenient. With *formatted input*, an *informat* follows each variable name in the **input** statement. The informat defines how a variable is to be read. A missing value is denoted by a period for numeric data and a blank for character data. There are many different types of informats for reading various types of data. We describe only the most commonly used informats and refer the reader to reference 1 in Appendix E otherwise.

w.d is for reading standard numeric data. The w represents the width in columns and the d value specifies the number of digits to the right of the decimal point. The numeric value can be anywhere in the field. If data already contain a decimal point, then d is ignored; otherwise the number is divided by 10^d. The w.d informat also reads numbers in scientific notation, e.g., $1.257E3 = 1.257 \times 10^3 = 1257$.

$CHARw. is for reading character data. The w value specifies the width in columns. Blanks are treated as characters with this informat. If you do not want to interpret leading or trailing blanks as characters, then use the informat $w., which is otherwise the same.

For example, suppose that the file C:\datafile.txt has two data lines in it. The first has the word michael in columns 2 to 8, the number 10 in columns 10, 11 and the number 1.3E5 in columns 13 to 17 and the second data line having the word evans in columns 4 to 8, the number 11 in columns 10, 11, and . in column 13. Then the program

```
data example;
infile 'datafile' pad;
input x $9.  y 2.0 @13 z 5.2;
cards;
```

creates two observations where the first has x = michael, y = 10, z = 130000 and the second has x = evans, y = 11, and z = .; i.e., z is missing in the second observation. Note that x occupies columns 1 to 9, y occupies columns 10 to 11, and z occupies columns 13 to 17.

A great convenience arises with formatted input when there are many variables that can be treated alike, for we can then use *grouped informat lists*. For this, we create lists of variables in parentheses, followed by a list of informats in parentheses. The informats are recycled until all variables are read. For example,

```
input (x y z) (3.  3.  3.);
```

reads three numeric variables, each occupying three columns, and is equivalent to

```
input (x y z) (3.);
```

as the informat 3. is recycled. If many variables are stored consecutively and all have the same length, then it is convenient to give them all the same first letter in their name. For example,

```
input (x1-x60) (1.)   #2 (x61-x70) (2.)   #3 x71 $4.;
```

reads in 60 variables x1, x2, ..., x60 stored consecutively in the first row, one column apiece, 10 variables x61, x62, ..., x70 stored consecutively in the second row, two columns apiece, and one variable x71, a character variable stored in the third row in four columns.

There are other forms of input and many other features of the ones discussed here. If you cannot input your data using one of the methods we have described, most probably an appropriate method exists. Consult reference 1 in Appendix E. For example, it is possible to define default informats for variables and specific informats for variables using the **informat** statement.

5.3.4.5 MISSING and LOSTCARD

Sometimes you want to specify that characters other than a period or a blank should be treated as missing data. For example, the **missing** statement in

```
data example;
missing a r;
input $ x $ y;
cards;
```

declares that the characters a and r are to be treated as missing values for character variables x and y whenever they are encountered in a record.

When multiple records are used to form a single observation, you must be careful that each observation has the correct number of records in the file. You can do this using the **lostcard** statement when each record contains a variable that is the same for each observation, i.e., an identifier. For example, suppose the file C:\records contains

```
100 1 2
100 3
101 4 5
102 6 7
102 8
```

Then the program

```
data example;
infile 'C:\records';
input id1 x y #2 id2 z;
if id1 ne id2 then
do;
put 'error in data records' 'id1 =' id1 'id2 =' id2;
lostcard;
end;
cards;
proc print;
```

checks to make sure that each observation has the right number of records and prints out an error message if they don't in the SAS Log window using the **put** statement described in Subsection I.5.4. The **lostcard** statement tells SAS to delete the observation where the error occurred and resynchronize the data. Hence, the data set **example** has two observations corresponding to id1 = id2 = 100 and id1 = id2 = 102 with three variables, x, y, and z. The **do-end** statements are discussed in Subsection I.5.5.

5.4 Output

5.4.1 Output to the Output Window

After we have input data we commonly want to print it to make sure that we have input the data correctly. In fact this is recommended. We have already seen that **proc print** can be used to print data records in the Output window. We discuss **proc print** more extensively in Subsection I.6.1. The output of most proc's are recorded in the Output window.

5.4.2 Output to the Log Window

The **put** statement can be used to write in the Log window or to an external file. To write in the Log window, the **put** statement occurs in a data step without a **file** statement. For example,

```
data example;
input x y;
z = x + y;
```

```
put 'x = ' x 'y = ' y 'z = ' z;
cards;
1 2
3 4
```

writes

```
x = 1 y = 2 z = 3
x = 3 y = 4 z = 7
```

in the Log window. Note that we can output characters by enclosing them in single quotes.

There are three basic types of output: *list, column,* and *formatted output.* These behave analogously to list, column, and formatted input. There is a pointer that controls the column where **put** prints a value with the same commands to control the pointer as in Subsection I.5.2. Rather than informats, we have *formats* with **put** statements. There are two basic formats, **w.d** and **$CHARw.** (and **$w.**), and they behave just as their corresponding informats do. For example,

```
data example;
input x y;
put x 5.3 @10 y 4.1;
cards;
1 2
3 4
```

writes
```
1.000 2.0
2.000 3.0
```

in the Log window. Note that x occupies columns 1 to 5 and y occupies columns 10 to13 and they are printed right-justified in their fields. There is also a **format** statement used to define default formats for variables and specific formats for variables or to define entirely new formats. See reference 1 in Appendix E for a discussion.

The **list** statement is used to list in the Log window the input lines for the observation being processed. This is useful for printing suspicious input lines. For example,

```
data example;
input x y z;
if y = .  then list;
cards;
```

prints out the entire observation each time the y variable is missing.

5.4.3 Output to a Text File

The **file** statement is used before **put** when we want a **put** statement to write to an external file. For example, if we want the output to go into a file called C:\records.txt, then the statement

```
file 'C:\records.txt';
```

must appear before the **put** statement. If C:\records.txt already exists, it will be overwritten, otherwise it is created.

5.4.4 Output to a SAS Data Set

As mentioned earlier, the data step works as an implicit loop. An observation is read in, then all the statements in the data step are executed, the observation is added to the SAS data set being created, and then the process starts all over again with a new observation until there are no more observations left to read. The **output** statement overrides this operation. For example, suppose we have a file c:\saslibrary\data.txt that contains observations with a character variable called **gender** and a numeric variable x. The variable **gender** takes the values M and F for male and female. The following program

```
data males females;
infile 'c:\saslibrary\data.txt';
input gender $ x;
if gender='M' then output males;
y=x**2;
cards;
```

creates two SAS data sets called **males** and **females**, where **males** contains all the observations with **gender** = M and **females** contains all the observations with **gender** = F. Note that an observation is written to the SAS data set named in the **output** statement as soon as the **output** statement is invoked. For example, in the program here the variable y has all missing values in `males` but takes the assigned value in `females`. If an **output** statement is used in a data step, then an **output** statement must be used whenever we want to output an observation to a SAS data set; SAS no longer implicitly outputs an observation at the end of the data step.

The output statement is very useful. For example, we can use it inside a **do-end** group to create variables and write them out as observations to a SAS data set even if the program does not input any observations. This is important, as we can then use SAS to carry out many computations for us. Even though we haven't discussed the **do** statement yet (see Subsection I.5.5), we illustrate with a simple example. The program

```
data one;
do i=1 to 101;
x=-3+(i-1)*(6/100);
y=1/(1+x**2);
output one;
end;
drop i;
cards;
```

creates a SAS data set `one` with 101 observations and two variables `x` and `y`. The variable `x` ranges from -3 to 3 in increments of $.06$ and $y = (1+x^2)^{-1}$. We use the **drop** statement to get rid of the index variable `i`. For example, we may wish to plot the points (x,y), and as we shall see, there are procs that permit us to do so.

5.4.5 Output to a SAS Data Set File

Sometimes we want to create a permanent record of a SAS data set. We do this by first creating a folder in which the files are to be stored. For example, we might create the folder `C:\saslibrary` and think of this as a library for various SAS data sets that we create and want to save. We can have a number of different *SAS libraries* that can hold data sets with common features. Suppose we have created a folder called `C:\saslibrary`. Then the **libname** statement in the program

```
libname storage 'C:\saslibrary';
data storage.example;
input y x1 x2;
cards;
```

```
1 2 3
4 5 6
run;
```

creates a permanent SAS file in the folder `C:\saslibrary` called `example.sd2` that contains a SAS data set with two observations and the three variables `y`, `x1`, and `x2`. This file contains a permanent record of the SAS data set `example`. This is useful when we have a very large data set, want to carry out many analyses on it, and don't want to run the data step every time we do an analysis.

A SAS data set stored in a file can be accessed by proc's for analysis. For example,

```
libname storage 'C:\saslibrary';
proc print data = storage.example;
```

prints out the contents of the SAS data set `example`.

If you should forget what a stored SAS data set contains, for example, the data set `example`, then use **proc contents** as in

```
libname storage 'C:\saslibrary';
proc contents data = storage.example;
```

It prints a listing in the Output window of the number of observations and the number of variables and their type in the SAS data set `example`.

Sometimes we want to alter the contents of a stored SAS data set, for example, change the values of some variables in some observations or add observations. To do so, we use the **update** statement. Suppose the SAS data set `one`, stored in the file `one.sd2` in folder `C:\saslibrary`, contains the variables `y`, `x1`, and `x2` and four observations with the values:

```
152 70 5000
160 66 4000
180 72 4500
240 68 7000
```

Then the program

```
libname storage 'measures';
data two;
input y x1 x2;
cards;
152 72 5000
```

```
232 60 2500
data storage.one;
update storage.one two;
by y;
```

produces a new data set **one** stored in `C:\saslibrary\one.sd2`;. the old file is overwritten, given by

```
152 72 5000
160 66 4000
180 72 4500
232 60 2500
240 68 7000
```

Note that the value of **x1** has been changed from 70 to 72 in the observation in the old data set corresponding to **y = 152** and a new observation has been added corresponding to **y = 232**. For **update** to work, both the old data set, called the *master data set,* in this case **one**, and the data set containing the changes, called the *transactions data set,* in this case **two**, must be sorted (see Subsection I.6.2) by a common variable, in this case **y**.

5.5 Control Statements

There are a number of statements that control the flow of execution of statements in the data step.

5.5.1 IF-THEN-ELSE

If-then-else statements are used to conditionally execute a SAS statement. They take the form

if *expression* **then** *statement1*;
else *statement2*;

where *statement1* is executed if *expression* is true, *statement2* otherwise. The **else** part is optional, and if it is left out, control passes to the first statement after the **if-then** statement when *expression* is false. For example,

```
data example;
input x $ y;
if x eq 'blue' then z = 1;
else z = 0;
```

```
cards;
red 1
blue 2
```

compares x to blue after input. A new variable, z, is set equal to 1 when x equals blue, otherwise z is set equal to 0. See Appendix A for a listing of all comparison and logical operators that can be used in such statements.

5.5.2 GOTO and RETURN

A **goto** statement tells SAS to jump immediately to another statement in the same data step and begin executing statements from that point. For example,

```
data info;
input x y;
if 1<=x then goto OK;
x = 3;
OK: return;
cards;
```

checks to see if the input value of x is greater than or equal to 1; if it is not, then x is set equal to 3; if it is then the SAS program jumps to the statement labelled OK. This is a **return** statement which tells SAS to begin processing a new observation.

5.5.3 STOP

The **stop** statement stops processing a SAS data step. The observation being processed when the **stop** statement is encountered is not added to the data set and processing resumes with the first statement after this data step. For example, in

```
data example;
input x y z;
if x = 2 then stop;
cards;
```

stops building the data set when a value of x = 2 is encountered.

5.5.4 DO-END

The **do** statement designates a group of statements to be executed as a unit until a matching **end** statement is encountered. A number of **do** statements can be nested within **do** groups. A simple **do** is often used within **if-then-else** statements to designate a group of statements to be executed depending on whether an if condition is true or false. For example, the program

```
data example;
input x;
if x gt 0 then
do;
y = x*x;
z = -x;
end;
else w = x;
cards;
```

creates two new variables **y** and **z** when **x** > 0 and one new variable **w** when **x** ≤ 0. Note that these variables are equal to . when they are not assigned anything.

There are several variations of the **do-end** statement. For example, the version that takes the form

do *index* $=$ *start* **to** *stop* **by** *increment*;
statements
end;

executes *statements* repetitively between **do** and **end** where *statements* is a group of SAS statements. This is called an *iterative do* statement. The number of times *statements* is executed is determined as follows. Initially the variable *index* is set at *start* and *statements* executed. Next the *increment* is added to the *index* and the new value is compared to *stop*. If the new value of *index* is less than or equal to *stop*, then *statements* is executed again; otherwise, it is not. If no *increment* is specified, the default is 1. The process continues until the value of *index* is greater than the value of *stop*, upon which control passes to the first statement past the **end** statement. For example, the program

```
data;
sum = 0;
do i=1 to 100;
sum = sum + i*i;
end;
put sum;
```

writes 338350 in the Log window. This is the sum $1^2 + 2^2 + \ldots + 100^2$.

We can also add **while** and **until** conditions to an iterative **do** statement, as in

do *index* = *start* **to** stop **by** *increment* **while** (*expression*);
do *index* = *start* **to** stop **by** *increment* **until** (*expression*);

For example,

```
data;
sum = 0;
do i = 1 to 100 until (sum gt 1.0E5);
sum = sum + i*i;
end;
put i sum;
```

writes 67 102510 in the Log window. The value 102510 is the sum $1^2 + 2^2 + \ldots + 67^2$. The **until** condition is checked after each iteration. If the condition is false, we stop iterating; otherwise we continue until *index* equals *stop*. A **while** condition is evaluated at the start of each iteration. If it is false, only that iteration and no further iterations are carried out. Otherwise the program continues iterating until *index* equals *stop*.

Another variant is the **do until** statement. For example, in

do until (*expression*);
statements
end;

statements is repetitively executed until *expression* is false. The value of *expression* is evaluated after each iteration. The **do while** statement is another variation, with **while** replacing **until**. In a **do while** statement, the expression is evaluated before each iteration, and the statements are executed while the expression is true. For example,

```
data;
sum = 0;
```

```
i = 1;
do while (i le 100);
sum = sum + i*i;
i = i+1;
end;
put sum;
```

writes 338350 in the Log window. This is the sum $1^2 + 2^2 + \ldots + 100^2$.

5.5.5 SELECT-OTHERWISE

The **select-otherwise** statement replaces a sequence of **if-then-else** statements. The **select** statement takes the form:

select (*expression*);
when (*expression1*) *statement1*;
when (*expression2*) *statement2*;

$\vdots$

otherwise *statement*;
end;

In this group of statements, SAS compares *expression* to *expressioni*. If they are equal, then *statementi* is executed. If none are equal, then *statement* is executed. The **otherwise** statement is optional. An **end** statement ends a **select** group. For example, the program

```
data example;
input x;
select(x);
when(1) z = 0;
when(2) z = 0;
when(3) z = 0;
otherwise z = 1;
end;
cards;
```

adds a variable z to the data set **example**, which takes the value 0 when x = 1, 2, or 3 and the value 1 when x takes any other value.

6 SAS Procedures

The analysis of a SAS data set takes place in proc steps. A proc step takes the form

proc *name* **data** = *dataset1* *options1*;
statements / options2

where *name* gives the name of the proc being used, **data** = *dataset1* gives the name of the SAS data set to be analyzed by the procedure (if it is omitted, the most recently created SAS data set is used), *options1* specifies other features specific to the proc, *statements* provides further instructions concerning the behavior of the proc with the data set, and *options2* specifies options for the way the *statements* work. Typically, there is a default setting for the *options*. For example, **data** = is an option, and if it is omitted the most recently created SAS data set is used. Another common option allows the proc to create a SAS data set as part of its output when this is appropriate.

There are many different proc's. For example, **proc reg** carries out a statistical analysis known as a regression analysis. Typically, we need to know more about the proc than just its name before we can use it effectively, as there are different options and statements that go with each one.

Many different proc's can be used on the same SAS data set and appear in the same SAS program. We discuss several important procedures in this section.

6.1 PROC PRINT

The procedure **proc print** lists data in a SAS data set as a table of observations by variables. The following statements can be used with **print**.

proc print *options*;
var *variables*;
id *variable*;
by *variables*;

The following option may appear in the **proc print** statement..

data = *SASdataset*

where *SASdataset* is the one printed. If none is specified, then the last SAS data set created is printed. If no **var** statement is included, then all variables in the data set are printed, otherwise only those listed, and in the order in which they are listed are printed. When an **id** statement is used, SAS prints each observation with the values of the **id** variables first rather than the observation number, which is the default. For example, if the output for an observation consists of several lines, each line starts with the values of the **id** variables for that observation. As a specific case, if data set **one** contains variables **name**, **x1-x100**, then

```
proc print data = one;
var x1-x50;
id name;
```

prints only variables **x1-x50**, and each line of output for an observation begins with the value of **name** for that observation. This provides an easy way of identifying all lines associated with observations on output listings when **name** is chosen evocatively.

The **by** statement is explained in Section I.6.2.

6.2 PROC SORT

The procedure **proc sort** sorts observations in a SAS data set by one or more variables, storing the resulting sorted observations in a new SAS data set or replacing the original. The following statements are used with **proc sort.**

proc sort *options*;
by *variables*;

Following are some of the options that may appear with the **proc sort** statement.

data = *SASdataset1*
out = *SASdataset2*

If the **out** = *SASdataset2* statement doesn't appear, then the SAS data set being sorted is overwritten.

A **by** statement *must* be used with **proc sort**. Any number of variables can be specified in the **by** statement. The procedure **proc sort** first arranges the observations in the order of the first variable in the **by** statement, then it sorts the observations with a given value of the first variable by the second variable, and so on. By order, we mean increasing in value, or ascending order. If we want a **by** variable to be used in descending order, then the word **descending** must precede the name of the variable in the **by** list. For example, suppose the data set **one** contains

```
variable x y   z
obs1      1 2 100
obs2      4 1 200
obs3      3 4 300
obs4      3 3 400
```

Then the statements

```
proc sort data=one out=two;
by x;
```

produce the data set **two**, which contains

```
variable x y   z
obs1      1 2 100
obs2      3 4 300
obs3      3 3 400
obs4      4 1 200
```

Notice that the values with a common value of x retain their relative positions from **one** to **two**. The statements

```
proc sort data=one out=two;
by descending x y;
```

produce the data set **two**, which contains

```
variable x y   z
obs1      4 1 200
obs2      3 3 400
obs3      3 4 300
obs4      1 2 100
```

In this case the observations are sorted first into descending order by x, then any observations that have a common value of x are sorted in ascending order by y. We can have any number of variables in the **by** statement with obvious generalizations for how the sorting procedure works.

We note that we have not restricted the **by** variables to be numeric and they needn't be. If a **by** variable is character, then the sorting is done using ASCII order. From smallest to largest, the ASCII sequence is

blank ! " # $ % & ' ( ) * + , - . / 0 1 2 ... 9 : ; < = > ? @ A B C ... Z [ ]
^_ ' a b c ... z { | }~

Notice that *A* is smaller than *a* here. The smallest value is a blank and the
largest is a ~. For numeric **by** variables, the order is the usual ordering of
real numbers, but a missing value is treated as being smaller than any other
real value.

Most proc's allow what is called *by group processing*. The data set being
used in the proc has typically been sorted by the **by** variables. The proc is
then applied to each of the subgroups specified by particular values of the
by variables. For example, if the data set **one** is as earlier, then

```
proc sort data=one out=two;
by x;
proc print data=two;
by x;
```

prints out the data in three groups identified by the value of x namely, the
values of y and z when x = 1, then the values of y and z when x = 3, and
finally, the values of y and z when x = 4. The statement by descending x
y; creates four groups. If the data are sorted in descending order for any of
the **by** variables, then the word **descending** must precede this variable in
the **by** statement of the proc.

In certain cases, as in the SAS data set **one**, the observations appear in
groups with a common value of a variable(s), in this case x, but they are not
sorted by this variable(s) in ascending or descending order. If we do not wish
to sort them, which can be time-consuming for large data sets, we can still
use by group processing, but the word **notsorted** must precede any such
variable(s) in the **by** statement of the proc.

7 Exercises

1. The following data give the high and low trading prices in Canadian dollars for various stocks on a given day on the Toronto Stock Exchange. Enter these data into a SAS data set with three variables, Stock, Hi, and Low and 10 observations. Print the data set in the Output window to check that you have successfully entered it. Save the data set as a permanent SAS data set giving it the name stocks.

Stock	Hi	Low
ACR	7.95	7.80
MGI	4.75	4.00
BLD	112.25	109.75
CFP	9.65	9.25
MAL	8.25	8.10
CM	45.90	45.30
AZC	1.99	1.93
CMW	20.00	19.00
AMZ	2.70	2.30
GAC	52.00	50.25

2. In a data step input the SAS data set stocks created in Exercise 1 from the file containing it. Calculate the average of the Hi and Low prices for all the stocks and save it in a variable called average. Calculate the average of all the Hi prices and output it. Do the same for all the Low prices. Save the data set using the same name. Write the data set stocks to a file called stocks.dat. Print the file stocks.dat on your system printer.

3. Using the SAS data set created in Exercise 2, calculate, using SAS commands, the number of stocks in the data set whose average is greater than $5.00 and less than or equal to $45.00.

4. Using the SAS data set created in Exercise 2, add the following stocks

to the permanent data set.

Stock	Hi	Low
CLV	1.85	1.78
SIL	34.00	34.00
AC	14.45	14.05

Remove the variable average from the data set and save the data set.

5. Using the data set created in Exercise 4 and the sort procedure, sort the stocks into alphabetical order. Save the data set.

6. Using the data set created in Exercise 5 and the by statement, calculate the average Hi price of all the stocks beginning in A.

7. Using the data set created in Exercise 5, recode all the Low prices in the range $0 to $9.99 as 1, in $10 to $39.99 as 2, and greater than or equal to $40 as 3, and save the recoded variable in a new variable.

Part II

SAS Procedures for Data Analysis

Chapter 1

Looking at Data: Distributions

SAS statements introduced in this chapter

options	proc gchart	proc plot	title
proc chart	proc gplot	proc timeplot	
proc freq	proc means	proc univariate	

This chapter of IPS is concerned with the various ways of presenting and summarizing a data set and also introduces the normal distribution. By presenting data we mean convenient and informative methods of conveying the information contained in a data set. There are two basic methods for presenting data, through graphics and through tabulations. It can be hard to summarize exactly what graphics or tables are saying about data. The chapter introduces various summary statistics that are commonly used to convey meaningful information in a concise way. The normal distribution is of great importance in the theory and application of statistics, and it is necessary to gain some facility with carrying out various computations with this distribution.

All these topics would involve much tedious, error-prone calculation if we did them by hand. In fact, you should almost never rely on hand calculation in carrying out a data analysis. Not only are there many far more important things for you to be thinking about, as the text discusses, but you are also likely to make errors. On the other hand, never blindly trust the computer! Check your results and make sure that they make sense in light of the application. For this a few simple hand calculations can prove valuable. In working through the problems in IPS you should try to use SAS as much as possible; your skill with the package will increase and inevitably your data analyses will be easier and more effective.

1.1   Tabulating and Summarizing Data

If a variable is categorical, we construct a table using the values of the variable and recording the *frequency* (count) and perhaps the *relative frequency* (proportion) of each value in the data. The relative frequencies serve as a convenient summarization of the data.

If the variable is quantitative, we typically *group* the data in some way, i.e., divide the range of the data into nonoverlapping intervals and then record the frequency and proportion of values in each interval. Grouping a variable is accomplished during the data step by introducing a new variable that takes a particular value whenever the original variable is in a certain range. If the original variable were `height` in feet, we might define a new variable `gheight` that takes the value 1 when `height` is less than 5, the value 2 when $5 \leq$ `height` < 6, and so on.

If the values of a variable are *ordered,* then we can record the *cumulative distribution*, the proportion of values less than or equal to each value. Quantitative variables are always ordered, but sometimes categorical variables are as well, e.g. when a categorical variable arises from grouping a quantitative variable.

Often it is convenient with quantitative variables to record the *empirical distribution function*, which for data values $x_1, \dots, x_n$ and at a value x is given by

$$\hat{F}(x) = \frac{\# \text{ of } x_i \leq x}{n}.$$

$\hat{F}(x)$ is the proportion of data values less than or equal to x. We can summarize such a presentation via the calculation of a few quantities such as the *first quartile*, the *median,* and the *third quartile* or the *mean* and the *standard deviation.*

1.1.1   PROC FREQ

The **proc freq** procedure produces frequency tables. A frequency table is table of counts of the values variables take. Frequency tables show the

FAMILY	Frequency	Percent	Cumulative Frequency	Cumulative Percent
1	2	20.0	2	20.0
2	3	30.0	5	50.0
3	2	20.0	7	70.0
4	1	10.0	8	80.0
5	1	10.0	9	90.0
6	1	10.0	10	100.0

Figure 1.1: Output from **proc freq**.

distribution of variable values and are primarily useful with variables where values are repeated in the data set. For example, suppose the data set one contains ten observations of the categorical variable family corresponding to the number of members in a family. The program

```
data one;
input family;
cards;
2
3
1
5
3
2
4
6
1
2
proc freq data=one;
tables family;
run;
```

produces the output shown in Figure 1.1, which gives the unique values taken by the variable family, the frequency or count of each variable value, the relative frequency or percent of each value, the cumulative frequency, and the cumulative relative frequency.

Some of the features of **proc freq** are described here. Other capabilities of this procedure are described in subsequent parts of this manual when we need them; in particular see Section II.2.2. Some of the statements used with **proc freq** follow.

proc freq *options*;
tables *requests/options*;
weight *variable*;
by *variables*;

Following are two options that can be used in the **proc freq** statement.

data = *SASdataset*
noprint

where *SASdataset* is the SAS data set containing the variables we want to tabulate.

Suppose the SAS data set **two** contains the variables a, b, c, and w. If you want a one-way frequency table for each variable, then simply name the variables in a **tables** statement. For example, the statements

```
proc freq data=one;
tables a b;
```

produce two one-way frequency tables, one giving the values of a and the number of observations corresponding to each value a assumes, and the other doing the same for b.

The commands

```
data;
input a;
cards;
1
1
0
0
1
1
proc freq noprint;
tables a/out=save;
proc print data=save;
run;
```

cause a SAS data set called **save** to be created that contains two observations, (corresponding to the number of values **a** assumes) and three variables **a**, count, and **percent** using the **out** = *SASdataset* option to the **tables** statement. The output

OBS	A	COUNT	PERCENT
1	0	2	33.3333
2	1	4	66.6667

is produced from **proc print data=save;**. Because of the **noprint** option to **proc freq**, this procedure yields no printed output.

If a **weight** statement appears, then each cell contains the total of all the *variable* values for the observations in that cell. For example,

```
data;
input w a;
cards;
1 0
2 1
3 0
4 1
-5 0
6 1
3.2 1
proc freq;
tables a;
weight w;
```

produces a 2 × 1 table with frequency -1 in the 0-cell and frequency 15.2 in the 1-cell.

Tables can also be produced for each by group as specified by *variables* in the **by** statement. See **proc sort** for further discussion.

1.1.2 Calculating the Empirical Distribution Function

We consider calculating the empirical distribution for Newcomb's measurements in Table 1.1 of IPS. Suppose these data are in the text file c:/saslibrary/newcomb.txt with each measurement on a single line. Then the program

LIGHT	Frequency	Percent	Cumulative Frequency	Cumulative Percent
-44	1	1.5	1	1.5
-2	1	1.5	2	3.0
16	2	3.0	4	6.1
19	1	1.5	5	7.6
20	1	1.5	6	9.1
21	2	3.0	8	12.1
22	2	3.0	10	15.2
23	3	4.5	13	19.7
24	5	7.6	18	27.3
25	5	7.6	23	34.8
26	5	7.6	28	42.4
27	6	9.1	34	51.5
28	7	10.6	41	62.1
29	5	7.6	46	69.7
30	3	4.5	49	74.2
31	2	3.0	51	77.3
32	5	7.6	56	84.8
33	2	3.0	58	87.9
34	1	1.5	59	89.4
36	4	6.1	63	95.5
37	1	1.5	64	97.0
39	1	1.5	65	98.5
40	1	1.5	66	100.0

Figure 1.2: Output from **proc freq** applied to the data set in Table 1.1 of IPS gives the empirical distribution function in the last column.

```
data newcomb;
infile 'c:/saslibrary/newcomb.txt';
input light;
cards;
proc freq data=newcomb;
tables light;
run;
```

reads the data from this file into the variable `light` in the SAS data set `newcomb` and then **proc freq** produces the table given as Figure 1.2. Note that the first column gives the unique values taken by the variable `light` and the last column gives the value of the empirical distribution function at each of these points. Note that from this printout we see that the value of the empirical distribution function at 16 is 6.1%.

1.1.3 PROC MEANS

Rather than printing out the entire empirical distribution function for a quantitative variable in a data set, it is sometimes preferable to record just a few numbers that summarize key features of the distribution. One possible choice

```
Mean and standard deviation for Newcomb data

Analysis Variable : LIGHT


  N          Mean        Std Dev         Minimum         Maximum
-------------------------------------------------------------------
 66     26.2121212     10.7453248     -44.0000000      40.0000000
-------------------------------------------------------------------
```

Figure 1.3: Output from running **proc means** on the Newcomb data.

is to record the mean and standard deviation of a variable. We can use the procedure **proc means** to do this in SAS. For example, suppose we want to use **proc means** on Newcomb's measurements in Table 1.1 of IPS, which is in the text file `c:/saslibrary/newcomb.txt`. Then the commands

```
options nodate nonumber;
data newcomb;
infile 'c:\saslibrary\newcomb.txt';
input light;
cards;
proc means;
title 'Mean and standard deviation for Newcomb data';
var light;
```

produce the output shown in Figure 1.3. This gives n the total number of observations used — observations for which the variable value is missing are not used — the mean, the standard deviation, the minimum data value, and the maximum data value.

We just introduced two new statements that can be used in SAS procedures. The **options** statement placed before a procedure affects the way the output from the procedure is written in the Output window. Here we have asked that the output suppress the date and the page number, both of which are usually printed on the output. There are many other options that can be specified; we refer the reader to reference 1 in Appendix E for a discussion. Also the **title** statement is used with **proc means** to write a title in the Output window in place of the default title "The SAS System". The **title** statement can be used with any procedure.

The procedure **proc means** has many features beyond just producing the output in Figure 1.3. Many other descriptive statistics can be calculated and saved permanently in a SAS data set. We describe here some of the more useful features of **proc means**. Spending some time now learning about the many features of **proc means** will be helpful in learning about the behavior of many of the procedures in SAS, as they have a number of features in common.

The statements that may appear with this procedure follow.

proc means *options keyword names*;
var *variables*;
class *variables*;
freq *variable*;
weight *variable*;
output out = *SASdataset* **keyword** = *names*;
by *variables*;

Following are some of the options that can be used in the **means** statement.

data = *SAS dataset*
nway
noprint

The following statistics may be requested with **proc means** by giving the *keyword names* of the statistics in the **proc means** statement. These keywords may also be used in the **output** statement. Some of the statistics listed here are not relevant to this part of the course, but we list them all for convenience later.

n number of observations on which the calculations are based.
nmiss number of missing values.
mean mean.
std standard deviation.
min smallest value.
max largest value.
range range = **max** − **min.**
sum sum.
var variance.
uss uncorrected sum of squares = sum of squares of the data values.
css corrected sum of squares = sum of squared deviations of the data values from their mean.

stderr standard error of the mean.
cv coefficient of variation (percent) = standard error divided by the mean and then multiplied by 100.
skewness measure of skewness.
kurtosis measure of kurtosis.
t Student's t value for testing the hypothesis that the population mean is 0.
prt probability of a greater absolute value of Student's t.

When no statistics are specifically requested in the **proc means** statement, the procedure prints only n, mean, standard deviation, minimum and maximum for each variable in the **var** statement. The results are printed in the order of the variables on the **var** statement. If no **var** statement is given then these statistics are computed for all numeric variables in the input SAS data set specified in the **data** option. If no data set is specified then the last SAS data set created in the program is used.

The variables in the **class** statement are used to form subgroups; i.e. these variables are used to classify observations. The **class** variables may be either numeric or character, but normally each variable takes a small number of values or levels. The **class** statement works like the **by** statement (see **proc sort**), but we do not require that the data set be sorted by the **class** variables as we do with the **by** statement. For example, suppose the SAS data set **one** contains

```
variable  x     y
obs1      1    1.1
obs2      1    2.2
obs3      2   -3.0
obs4      2    2.0
```

Then the statements

```
proc means data=one mean std cv;
class x;
```

form two *class groups*; when x = 1 and x = 2, and compute the mean, standard deviation and coefficient of variation for each group. The output from this program is given in Figure 1.4.

If the **freq** *variable*; statement appears, then the i-th observation occurs in the original data set the number of times given by the i-th value of *variable*. Therefore the values assumed by *variable* must be positive integers or that observation is not included in the calculations. For example, if the i-th

```
Analysis Variable : Y
```

X	N Obs	Mean	Std Dev	CV
1	2	1.6500000	0.7778175	47.1404521
2	2	-0.5000000	3.5355339	-707.1067812

Figure 1.4: Means, standard deviations and coefficients of variation for two class groups.

observation occurred f_i times in the original data, then the mean of a variable x is

$$\overline{x} = \frac{\sum_{i=1}^{n} f_i x_i}{\sum_{i=1}^{n} f_i}$$

and the i-th value of *variable* must be equal to f_i.

The **weight** *variable* statement behaves similarly to the **freq** *variable* statement. In this case the i-th value of variable is a weight $w_i \geq 0$ that is applied to the i-th observation in calculations. For example, the weighted mean is given by

$$\overline{x} = \frac{\sum_{i=1}^{n} w_i x_i}{\sum_{i=1}^{n} w_i}.$$

Note that the weights are not constrained to be nonnegative integers.

The **output** statement requests that **proc means** output statistics to a new SAS data set that is specified as *SASdataset* in **out** = *SASdataset*. The list of statistics specifies which statistics are to be included in the output data set. The *names* are used to give names to the output variables. For example, if data set **one** contains variables **x**, **y**, then

```
proc means data=one;
var x y;
output out=two mean=mn1 mn2 std=std1 std2;
```

```
                      The SAS System
   OBS     _TYPE_     _FREQ_    MN1     MN2      STD1      STD2

    1         0         4       1.5    0.575   0.57735   2.43088
```

Figure 1.5: Data set created by output statement consisting of values of statistics computed on primary data.

```
proc print data=two;
```

creates a SAS data set two with one observation and variables mn1, mn2, std1, and std2, where mn1 is the mean of x, mn2 is the mean of y, std1 is the standard deviation of x, and std2 is the standard deviation of y. Therefore the first name in a list is associated with the value of the statistic for the first variable in the **var** list, the second name is associated with the value of the statistic for the second variable in the **var** list, and so on. The output from printing the data set two is given in Figure 1.5. The value of putting these statistics in a SAS data set is that they are now available for analysis by other SAS procedures.

Notice that two additional automatic variables are added to the output data set, namely, _type_ and _freq_. The value of _freq_ is the number of data values used to compute the value of the statistic. If there is no **class** statement, then _type_ always equals 0 and there is only one observation in the output data set. When class variables are included, SAS creates a single observation for each subgroup and adds variables for the class variables taking the values that define the subgroups. In addition, observations that depend on the class variables are added. The values of _type_ are more complicated to explain, while _freq_ retains its earlier interpretation. To suppress the addition of these observations, which is what we typically want to do, use the **nway** option in the **proc means** statement. For example, if the SAS data set **three** contains

```
variable a b x  y
obs1    -1 1 6  0
obs2     1 2 3 -1
obs3    -1 1 2  2
```

```
obs4      1 2 0  3
obs5     -1 2 1  4
obs6      1 2 3  3
```

then

```
proc means data=three nway;
class a b;
var x y;
output out=four means=mnx mny;
```

creates the SAS data set `four` which contains

variable	a	b	type	freq	mnx	mny
obs1	-1	1	3	2	4	1.00000
obs2	-1	2	3	1	1	4.00000
obs3	1	2	3	3	2	1.66667

Therefore there are three unique values of (a,b). The value $(-1, 1)$ has `_freq_=2` observations, and the means of the x and y values for that subgroup are `mnx = 4` and `mny = 1`, respectively. For more details on the `_type_` variable, see reference [2] in Appendix E. Similar considerations apply when a **by** statement is included.

If we do not want **proc means** to print any output, e.g. when we want only to create an output data set, then we use the **noprint** option in the **proc means** statement.

1.1.4   PROC UNIVARIATE

The **proc univariate** procedure is used to produce descriptive summary statistics for quantitative variables. It is similar to **proc means**, but it has some unique features. For example, **proc means** permits class variables while **proc univariate** does not. On the other hand **proc univariate** allows for many more descriptive statistics to be calculated. Features in **proc univariate** include detail on the extreme values of a variable, quantiles, several plots to picture the distribution, frequency tables, and a test that the data are normally distributed. For example, suppose we want to use **proc univariate** on Newcomb's measurements in Table 1.1 of IPS, which is in the text file `c:/saslibrary/newcomb.txt`. Then the commands

```
options nodate nonumber;
```

Summary of Newcomb Data

Univariate Procedure

Variable=LIGHT

		Moments				Quantiles(Def=5)			
N	66	Sum Wgts	66	100% Max	40	99%	40
Mean	26.21212	Sum	1730	75% Q3	31	95%	36
Std Dev	10.74532	Variance	115.462	50% Med	27	90%	36
Skewness	-4.59848	Kurtosis	28.61494	25% Q1	24	10%	21
USS	52852	CSS	7505.03	0% Min	-44	5%	16
CV	40.99372	Std Mean	1.322658			1%	-44
T:Mean=0	19.81776	Pr>	T		0.0001	Range	84		
Num ^= 0	66	Num > 0	64	Q3-Q1	7		
M(Sign)	31	Pr>=	M		0.0001	Mode	28		
Sgn Rank	1038.5	Pr>=	S		0.0001				

Extremes

Lowest	Obs	Highest	Obs
-44(	6)	36(	37)
-2(	10)	36(	41)
16(	65)	37(	31)
16(	8)	39(	55)
19(	20)	40(	9)

Figure 1.6: Output from **proc univariate** applied to Newcomb's data.

```
data newcomb;
infile 'c:/saslibrary/newcomb.txt';
input light;
cards;
proc univariate;
title 'Summary of Newcomb Data';
var light;
run;
```

produce the output shown in Figure 1.6.

We observe from Figure 1.6 that running **proc univariate** has resulted in many summary statistics being printed for the data. We do not try to explain them all here but note that the total number n of observations on which the calculations are based is printed as well as the mean, standard deviation, median, quartiles, interquartile range, minimum, and maximum. So a great deal of information about the distribution of the variable can be obtained from **proc univariate**.

The following statements can be used with **proc univariate**.

proc univariate *options*;
var *variables*;
freq *variable*;

weight *variable*;
id *variables*;
output out=*SASdataset statistic-keyword*=*names*;
by *variables*;

Some of the options that can appear in the **univariate** statement are

data = *Sasdataset*
plot
freq
normal
noprint

The **by**, **freq**, and **weight** statements work as in **proc means**. The **output** statement is used when you want to create an output SAS data set with the values of statistics specified as the values of the variables. The new data set contains an observation for each **by** group specified. For example, if the SAS data set one contains the variables a, x, y, and z and it has been sorted by a, then

```
proc univariate data=one;
var x y z;
by a;
output out=two mean=mnx mny mnz std=stdx stdy stdz;
```

creates a SAS data set two which has seven variables a, mnx, mny, mnz, stdx, stdy, and stdz, with an observation for each value assumed by a. For a value of a, the variable a in the new data set takes this value, mnx is the mean of the x values corresponding to this value of a, and so on. If we use the **noprint** option, then SAS only creates the data set. The statistics that follow may be saved by **proc univariate** by giving the *keyword names* of the statistics in the **output** statement. Note that the statistics include all those available in **proc means**.

n
nmiss number of missing observation
nobs number of observations, **nobs** = **n** + **nmiss**
mean
stdmean standard deviation of the mean
sum
std

var

cv

uss

css

skewness

kurtosis

sumwgt

min

max

range

q3 3rd quartile or 75th percentile

median

q1 1st quartile or 25th percentile

qrange interquartile range, **q3 − q1**

p1 1st percentile

p5 5th percentile

p10 10th percentile

p90 90th percentile

p95 95th percentile

p99 99th percentile

mode

t Student's t statistic

probt probability of greater absolute value for Student's t

msign sign statistic

probm probability of greater absolute value for the sign statistic

signrank Wilcoxon statistic

probs probability of a greater absolute value for the centered Wilcoxon statistic

normal Shapiro-Wilk test statistic for testing normality when sample size is less than 200, otherwise the Kolmogorov statistic

probn P-value for testing the hypothesis that the sample is from a normal distribution

If the **plot** option is used, then a stem-and-leaf plot, a boxplot, and a normal probability plot are produced for each variable. The **freq** option requests a frequency table consisting of the variable values, frequencies, percentages, and cumulative percentages. The **normal** option causes a test statistic to be computed that tests the null hypothesis that the sample comes from a normal distribution.

1.2   Graphing Data

One of the most informative ways of presenting data is via a graph. There are several methods for obtaining graphs in SAS. For example, suppose we use **proc univariate** on Newcomb's measurements in Table 1.1 of IPS, which is in the text file `c:/saslibrary/newcomb.txt`. Then the commands

```
data one;
infile 'c:\saslibrary\newcomb.txt';
input x;
proc univariate plot;
var x;
run;
```

produce the usual output from this procedure together with a *stem-and-leaf plot*, a *boxplot*, and a *normal probability plot* for this data. In Figure 1.7 the stem-and-leaf and the boxplot are shown. For the boxplot the central horizontal line is the median with the next two lines the quartiles. The vertical lines extending from the quartiles extend to the minimum of the range of the data or 1.5 times the interquartile range, whichever is less. Any values further from the median than this but less than three interquartile ranges are marked with a 0, while more extreme values are marked with an asterisk. The plus sign represents the sample mean.

1.2.1   PROC CHART

The procedure **proc chart** produces *vertical and horizontal bar charts*, *block charts*, and *pie charts*. These charts are useful for showing pictorially a variable's values or the relationships between two or more variables. The following statements can be used with this procedure.

proc chart *options*;
vbar *variables/options*;
hbar *variables/options*;
block *variables/options*;
pie *variables/options*;

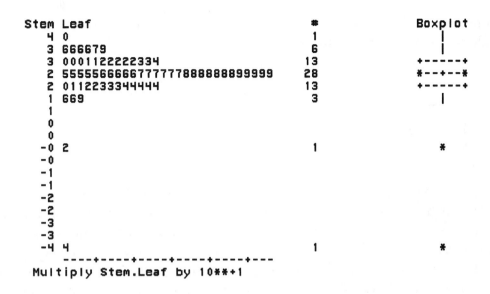

```
Stem Leaf                                       #       Boxplot
   4 0                                           1          |
   3 666679                                      6          |
   3 0001122222334                              13       +------+
   2 5555566666777777888888899999              28       *--+--*
   2 0112233344444                             13       +------+
   1 669                                        3          |
   1
   0
   0
  -0 2                                          1          *
  -0
  -1
  -1
  -2
  -2
  -3
  -3
  -4 4                                          1          *
     ----+----+----+----+----+---
Multiply Stem.Leaf by 10**+1
```

Figure 1.7: Stem-and-leaf plot and boxplot for Newcomb's data.

by *variables*;

An option available in the **proc chart** statement is

data = *SASdataset*

where *SASdataset* specifies the SAS data set containing the variables you want to plot.

In the **vbar** statement, list the variables for which you want *vertical bar charts*. Each chart takes one page. The **hbar** statement requests a horizontal bar chart for each variable listed. Each chart occupies one or more pages. In the **block** statement, list the variables for which you want *block charts*. See reference 2 in Appendix E for further discussion of block charts. The **pie** statement requests a *pie chart* for each variable listed. Each pie chart takes one page. See reference 2 in Appendix E for further discussion of pie charts.

The following options may be used with the **vbar** and **hbar** statements. If they are used, a slash (/) must precede the option keywords.

discrete Used when the quantitative variable specified is discrete. If **discrete** is omitted, then **proc chart** assumes that all numeric variables are continuous.

type = Specifies what the bars or sections in the chart represent. The following statements are used:

```
type = freq;
type = pct;
type = cfreq;
type = cpct;
type = sum;
type = mean;
```

The abbreviation **freq** makes each bar or section represent the frequency with which a value occurs for the variable in the data, **pct** makes each bar or section represent the percentage of observations of the variable having a given value, **cfreq** makes each bar or section represent cumulative frequency, **cpct** makes each bar or section represent cumulative percentage, **sum** makes each bar or section represent the sum of the **sumvar** = *variable* for observations having the bar's value, and **mean** makes each bar or section represent the mean of the **sumvar** = *variable* for observations having the bar's value. If no **type** = is specified, the default is **type** = **freq**.

sumvar = *variable* names the variable to collect summaries for means, sums or frequencies.

group = *variable* produces side-by-side charts with each chart representing the observations having a given value of *variable*. This variable can be numeric or character and it is assumed to be discrete.

subgroup = *variable* subdivides each bar into characters that show *variable*'s contribution to the bar.

For example, suppose we use **proc chart** on Newcomb's measurements in Table 1.1 of IPS, which has been placed in the SAS data set **one** with the measurements in the variable **x**. Then the commands

```
proc chart data=one;
vbar x/ type=pct;
run;
```

produce a vertical bar chart as presented in Figure 1.8. In this case the vertical bar chart is also called a *histogram* of **x**. SAS sorts the **x** values, divides them into subgroups according to whether or not they fall into equal-length subintervals, and then plots the proportion of the total number of values falling into that subinterval as the height of a vertical bar over the midpoint

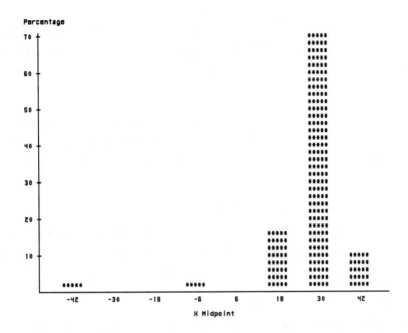

Figure 1.8: Histogram of Newcomb's data.

value for the subinterval. SAS has an internal algorithm for choosing the various parameters that determine the appearance of the plot.

As another example of plotting a histogram, suppose the SAS data set **two** contains a numeric variable **x** and a character variable **sex**, which takes the values M and F. Then

```
proc chart data=two;
hbar sex;
```

produces a horizontal bar chart, with the length of the two bars representing the frequencies, or counts. The commands

```
proc chart data=two;
vbar x / subgroup=sex;
```

produce a vertical bar chart for **x** with each bar divided to show how many M and F are in the subgroup.

If you do not like the appearance of the plot automatically produced by the algorithm in **proc chart**, you can change the character of the plot using the following options.

midpoints = *values* defines the range of values each bar or section represents by specifying the range midpoints. This is the most important option in determining the appearance of the plot. For example, the statement

```
vbar x / midpoints=10 20 30 40 50;
```

produces a chart with five bars: the first bar represents the range of data values with a midpoint of 10, the second bar represents the range of data values with a midpoint of 20, and so on. You can also abbreviate the list of midpoints as

```
vbar x/ midpoints=10 to 50 by 10;
```

which produces the same result. For character variables, **midpoints** may be useful in specifying a subset of the possible values. For example, you can give a list of the form

```
vbar sex / midpoints=m;
```

which produces a bar chart with only one bar for **sex**, giving the frequency of M.

axis = *value* specifies the maximum value to use in constructing the **freq**, **pct**, **cfreq**, or **cpct** axis.

The following options may be used with the **vbar** and **hbar** statements.

levels = *n* specifies the number of bars *n* representing each variable when the variables given in the **vbar** statement are continuous.

symbol =*'char'* defines the symbol *char* to be used in the body of standard **hbar** and **vbar** charts. The default **symbol** value is the asterisk '*'.

missing specifies that missing values are to be considered as valid levels for the chart variable.

nozeros specifies that any bar with zero value be suppressed.

ascending prints the bars in ascending order of size within groups.

descending prints the bars in descending order of size within groups.

The following options may be used with the **hbar** statement.

nostat specifies that no statistics be printed with a horizontal bar chart.

freq specifies that the frequency of each bar be printed to the side of the chart.

cfreq specifies that the cumulative frequency be printed.

percent specifies that the percentages of observations having a given value for the chart variable be printed.

cpercent specifies that the cumulative percentages be printed.

sum specifies that the total number of observations that each bar represents be printed.

mean specifies that the mean of the observations represented by each bar be printed.

For charts produced with any **type** = specification without a **sumvar** = *variable* option, **proc chart** can print **freq, cfreq, percent,** and **cpercent**. For **type** = **mean** with a **sumvar**= *variable* option, **proc chart** can print **freq** and **mean**. For a **type** = **sum** specification, **proc chart** can print **freq** and **sum**.

By group processing is also available with **proc chart**. For a discussion, see **proc sort**.

1.2.2 PROC TIMEPLOT

The procedure **proc timeplot** is used to plot time series. For example, the program

```
data one;
input group y z;
cards;
1 2.2 5.0
1 4.2 4.5
1 3.3 2.3
1 4.0 1.1
1 5.0 2.1
2 2.6 5.1
2 3.3 4.2
2 5.3 3.2
2 6.1 2.4
2 1.5 3.2
proc timeplot data=one uniform;
plot y = '*' z ='+' /overlay;
by group;
run;
```

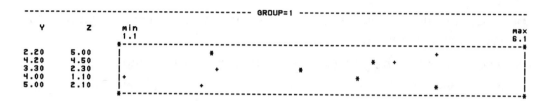

Figure 1.9: Overlaid time plots from **proc timeplot**.

creates two plots, one for each **by** group as specified by the value of the variable **group**. The **uniform** option to **proc timeplot** specifies that the horizontal scale, which is the scale for the variables, is the same for each plot and a time plot is given for variables y and z with these plots overlaid as specified by the **overlay** option to the **plot** statement. The time plot for **group=1** is given in Figure 1.9. Notice that the vertical axis is time, and of course we are assuming here that the order in each group corresponds to the time order in which each observation was collected. Also the values of the variables are printed by the time axis. In the **plot** statement we list all the variables for which time plots are requested and here specify the plotting symbols, as in y = '*' z ='+', which indicates that each value of y is plotted with * and each value of z is plotted with +. As with **proc means**, a **class** statement can also be used with **proc timeplot**.

1.3   Graphing Using SAS/Graph

Higher-resolution plots than those discussed for **proc chart** are available if you have the SAS/Graph software as part of your SAS implementation. SAS/Graph plots are displayed in the *Graph window,* which can be accessed via Globals ▶ Graph. Not only are there a number of much more elaborate plots, but plots can be enhanced in many ways, e.g., through the use of color. It is possible to edit a graph using Edit ▶ Edit graph, e.g., typing text directly onto the graph. The graph can be printed using File ▶ Print. Also the plot can be saved in some format such as .bmp or .jpeg using Export ▶, and graphics files in different formats can be imported into SAS/Graph using Import ▶ .

We can describe only a small proportion of the features available in

SAS/Graph and refer the reader to references 5 and 6 in Appendix E. Hereafter we will primarily present SAS/Graph plots because of their better appearance.

1.3.1   PROC GCHART

The procedure **proc gchart** works just like **proc chart** but it plots to the Graph window, the plots have a more professional look, and there is much more control over the appearance of the plot. For example, suppose we use **proc gchart** on Newcomb's measurements in Table 1.1 of IPS, which has been placed in the SAS data set **one** with the measurements in the variable x. Then the commands

```
axis1 label=('Passage time of light') length=4 in;
axis2 length=4 in;
proc gchart data=one;
vbar x / midpoints=-40 to 40 by 10 frame maxis=axis1
        raxis=axis2;
run;
```

produce Figure 1.10, which is a frequency histogram of the data. Here we have used the **axis** statement to define two axes (there can be up to 99 different axes defined); axis1 has the label Passage time of light and the length 4 inches (can also use centimeters) and axis2 has no label but is required to have a length of 4 inches as well. We assigned axis1 to the midpoint axis **maxis** and axis2 to the response axis **raxis** as options in the **vbar** statement. The **frame** option in the **vbar** statement ensures that a frame is drawn around the plot. As with any proc, we can also place a title on the plot if we wish by using a **title** statement. There are many additional ways in which the appearance of this plot can be controlled; we refer the reader to reference 6 in Appendix E.

1.4   Normal Distribution

It is important in statistics to be able to do computations with the normal distribution. As noted in IPS, the equation of the *density curve* for the

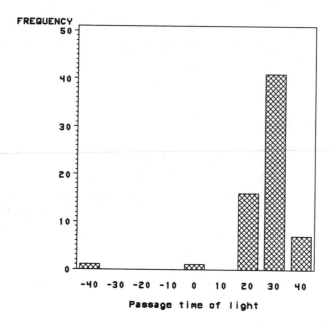

Figure 1.10: Histogram of Newcomb's data.

normal distribution with mean μ and standard deviation σ is given by

$$\frac{1}{\sqrt{2\pi}} e^{-\frac{1}{2}\left(\frac{z-\mu}{\sigma}\right)^2}$$

where z is a number. We refer to this as the $N(\mu, \sigma)$ density curve. Also of interest is the area under the density curve from $-\infty$ to a number x, i.e., the area between the graph of the $N(\mu, \sigma)$ density curve and the interval $(-\infty, x]$. Recall that this is the value of the $N(\mu, \sigma)$ distribution function at x and it equals the value of the $N(0, 1)$ distribution function at the *standardized value*

$$z = \frac{x - \mu}{\sigma}.$$

The $N(0, 1)$ distribution function is available in SAS as **probnorm** (see Appendix A). Sometimes we specify a value p between 0 and 1 and then want to find the point x_p such that p of the area for the $N(\mu, \sigma)$ density curve under its graph over $(-\infty, x_p]$. The point x_p is called the *p-th percentile* of the $N(\mu, \sigma)$ density curve. If z_p is the p-th percentile of the $N(0, 1)$ distribution, then $x_p = \mu + \sigma z_p$. The function **probit** is available to evaluate percentiles for the $N(0, 1)$ distribution.

All these calculations can be carried out from within the data step in SAS. For example, if $\mu = 5, \sigma = 2.2, p = .75$, and $x = 7.3$, then the program

```
data;
mu=5;
sigma=2.2;
x=7.3;
z=(x-mu)/sigma;
put 'The standardized value = ' z;
pn=probnorm(z);
put 'The N(5,2.2) distribution function at 7.3 = ' pn;
ip=probit(.75);
ip=mu+sigma*ip;
put 'The 75-th percentile of the N(5,2.2) distribution = '
        ip;
cards;
run;
```

writes

```
The standardized value = 1.0454545455
The N(5,2.2) distribution function at 7.3 = 0.8520935309
The 75-th percentile of the N(5,2.2) distribution
    = 6.4838774504
```

in the Log window.

1.4.1 Normal Quantile Plots

Some statistical procedures require that we assume that values for some variables are a sample from a normal distribution. A *normal quantile plot* is a *diagnostic* that checks for the reasonableness of this assumption. Note that *quantile* means the same as percentile. To create such a plot we use **proc univariate** with the **plot** option. For example, if the Newcomb data is in the data set **one** as the single variable **light**, then the commands

```
proc univariate plot;
var light;
```

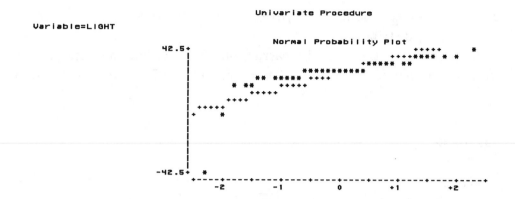

Figure 1.11: Normal quantile plot of Newcomb's data.

produce as part of the output the normal quantile plot shown in Figure 1.11. Note that in this plot asterisks * correspond to data values and pluses + correspond to a reference straight line. We see in this example a marked deviation from a straight line. See IPS for further discussion of this example.

1.5 Exercises

When the data for an exercise come from an exercise in IPS, the IPS exercise number is given in parentheses ( ). All computations in these exercises are to be carried out using SAS, and the exercises are designed to reinforce your understanding of the SAS material in this chapter. More generally, you should use SAS to do all the computations and plotting required for the problems in IPS.

1. Using Newcomb's measurements in Table 1.1 of IPS, create a new variable by grouping these values into three subintervals $(-50, 0)$, $[0, 20)$, $[20, 50)$. Calculate the frequency distribution, the relative frequency distribution, and the cumulative distribution of this ordered categorical variable.

2. Using the data in Example 1.5 of IPS on the amount of money spent by shoppers in a supermarket, print the empirical distribution function

and determine the first quartile, median, and third quartile. Also use the empirical distribution function to compute the 10-th and 90-th percentiles.

3. (1.23) Use SAS commands for the stem-and-leaf plot and the histogram. Use SAS commands to compute a numerical summary of this data and justify your choices.

4. (1.24) Transform the data in this problem by subtracting 5 from each value and then multiplying by 10. Calculate the means and standard deviations, using any SAS commands, of both the original and transformed data. Compute the ratio of the standard deviation of the transformed data to the standard deviation of the original data. Comment on this value.

5. (1.27) Transform this data by multiplying each value by 3. Compute the ratio of the standard deviation to the mean (called the *coefficient of variation*) for the original data and for the transformed data. Justify the outcome.

6. (1.38) Use SAS to draw time plots for the mens and women's winning times in the Boston marathon on a common set of axes.

7. For the $N(6, 1.1)$ density curve, compute the area between the interval (3,5) and the density curve. What number has 53% of the area to the left of it for this density curve?

8. Use SAS commands to verify the 68–95–99.7 rule for the $N(2, 3)$ density curve.

9. Use SAS commands to make the normal quantile plot presented in Figure 1.32 of IPS.

Chapter 2

Looking at Data: Relationships

SAS statements introduced in this chapter

proc corr   proc gplot   proc plot   proc reg   proc tabulate

In this chapter we describe SAS procedures that permit the analysis of relationships among two variables. The methods are different depending on whether or not both variables are quantitative, both variables are categorical, or one variable is quantitative and the other is categorical. Graphical methods are useful in looking for relationships among variables, and we examine various plots.

2.1 Relationships Between Two Quantitative Variables

2.1.1 PROC PLOT and PROC GPLOT

A *scatterplot* of two quantitative variables is a very useful technique when looking for a relationship between two variables. By a scatterplot we mean a plot of one variable on the y axis against the other variable on the x axis. For example, consider Example 2.4 in IPS where we are concerned with the relationship between the length of the femur and the length of the humerus for an extinct species. Then the program

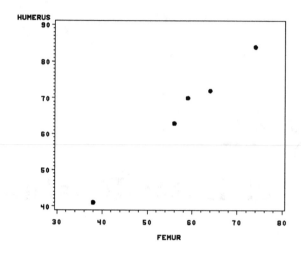

Figure 2.1: Scatterplot of humerus length versus femur length in Example 2.4 of IPS using **proc gplot**.

```
data archaeop;
input femur humerus;
cards;
38 41
56 63
59 70
64 72
74 84
symbol value=dot;
axis1 length=4 in;
axis2 length=5 in;
proc gplot data=archaeop;
plot humerus*femur/ vaxis= axis1 haxis = axis2 frame;
run;
```

produces the high-resolution scatterplot shown in Figure 2.1. This program uses the SAS/Graph procedure **proc gplot**, but if it is not available you can substitute **proc plot**. We describe **proc plot** first and then **proc gplot**.

PROC PLOT

The **proc plot** procedure graphs one variable against another, producing a scatterplot. This procedure takes the values that occur for each observation in an input SAS data set on two variables, say x and y, and plots the values of (x,y), one for each observation. The following statements are used with **proc plot**.

proc plot *options*;
plot *requests/options*;
by *variables*;

Following are some of the options available in the **proc plot**.

data = *SASdataset*
uniform

Here *SASdataset* represents the SAS data set that contains the variables to be plotted.

The **plot** statement lists the plots to be produced. You may include many **plot** statements and specify many *plot requests* on one **plot** statement. The general form of the plot request is *vertical * horizontal*;. first you name the variable to be plotted on the y axis, then a *, and then the variable to be plotted on the x axis. When a point on the plot represents a single observation, the letter A is used to represent this point. When a point represents two observations, the letter B is used, and so on. When a value of a variable is missing, that point is not included in the plot.

Another form of the **plot** request is *vertical*horizontal='character'*. With this form, you are naming the variables to be plotted on the y and x axes and also specifying a character (inside single quotation marks) to be used to mark each point on the plot.

A further form of the **plot** request is *vertical*horizontal=variable*, where the value of *variable* is now printed to mark each point. If you want to plot all combinations of one set of variables with another, you can use a grouping specification. For example, if the SAS data set **one** contains variables x1, x2, y1, y2, y3, then the statement

```
plot (x1 x2) * (y1-y3);
```

is equivalent to

```
plot x1*y1 x1*y2 x1*y3 x2*y1 x2*y2 x2*y3;
```

and produces six plots. If a variable appears in both lists, then it will not be plotted against itself. To plot all unique combinations of a list of variables, simply omit the second list. For example,

```
plot (x1 - x3);
```

produces plots **x1*x2**, **x1*x3**, **x2*x3**. If a **by** *variables*; statement is included, then the plot requests are carried out for each by group of observations. When the **uniform** option is specified, all the axes have the same scale for each pair of variables and for each by group so that the plots are comparable.

The **plot** statement also has a number of options. The options are specified by placing a slash / after the plot requests and then listing the options. For example,

```
plot x1*x2='+' x1*x3='.'  / overlay;
```

causes the **x1*x2** and **x1*x3** plots to be overlaid on the same set of axes with different plotting characters for each plot. It is possible to control the appearance of the axes using the **haxis** and **vaxis** options. For example,

```
plot x1*x2 / haxis = 10 20 30 40 vaxis = -3 -2 -1 0 1 2 3;
```

plots **x1** against **x2** with the tick marks on the x axis at 10, 20, 30 and 40 and the tick marks on the y axis at -3, -2, -1 0, 1, 2 and 3. Alternatively, as we have equispaced tick marks, we can write this command as

```
plot x1*x2 / haxis = 10 to 40 by 10 vaxis = -3 to 3 by 1;
```

produce the same graph..

There are many other options for the **plot** statement. For example, we can control the size of the plots so that multiple plots can appear on one page. See reference 2 in Appendix E for a full discussion of **proc plot**.

PROC GPLOT

The procedure **proc gplot** produces higher-resolution graphics such as Figure 2.1. We have much more control over the appearance of the plot with this procedure. The following statements are among those that can be used with this procedure.

proc gplot *options*;
plot *requests/options*;
by *variables*;

As with **proc gchart** we can also use **axis** statements (up to 99, labeled **axis1**, **axis2**, and so on) to control the appearance of the axes. Alternatively, we can control the appearance of the axes as in **proc plot**. For example, the statements

```
symbol value=dot;
axis1 30 to 80 by 10;
axis2 40 to 90 by 5;
proc gplot data=archaeop;
plot humerus*femur/ haxis = axis1 vaxis = axis2;
run;
```

substituted into the program that produced Figure 2.1 produce the same graph.

The default plotting symbol is +. The **symbol** statement is used to define alternative plotting characters and control other characteristics such as whether or not we want to join the points. More than one **symbol** statement can appear (up to 99, labeled **symbol1**, **symbol2**, and so on) When more than one plot is requested, the appropriate symbol statement is referenced by using *vertical*horizontal=n* where *n* refers to the *n*-th symbol generated in the program (not necessarily the one labeled this). Actually a full description of the use of the **symbol** statement is reasonably complicated, particularly if colors are being used, and we refer the reader to references 5 and 6 in Appendix E. It is better to specify the color of each symbol if we want to ensure that *n* refers to the *n*-th **symbol** statement. For example, the program

```
data one;
input x y z;
cards;
```

```
1 3.2 4.3
2 2.1 1.0
3 6.3 2.1
4 4.3 3.3
5 1.0 0.0
axis1 length=4 in label= ('independent variable');
axis2 length=4 in label =(' dependent variables');
symbol1 value=dot interpol=join color=black;
symbol2 value=circle color=black;
proc gplot data=one;
plot y*x=1 z*x=2/overlay haxis = axis1 vaxis =axis2 legend;
run;
```

produces Figure 2.2. Notice that we have specified the color for both plotting symbols, as this ensures that symbol1 corresponds to y*x and symbol2 corresponds to z*x. Note that **value** = *symbol* specifies the plotting character used, and some of the possibilities are

plus +
x ×
square □
dot •
circle ○
diamond ◇
triangle △

Many more are listed in reference 5 of Appendix E. Also we joined the points in the first plot of Figure 2.2 using **interpol** in symbol1 and requested a legend for the graph using the **legend** option in the **plot** statement.

There are many more features for **proc gplot**. We will encounter some of them in the remainder of this manual. In particular, if we want a boxplot of a quantitative variable, we use interpol=box in the relevant **symbol** statement.

2.1.2   PROC CORR

While a scatterplot is a convenient graphical method for assessing whether or not there is any relationship between two variables, we would also like to assess their relationship numerically. The *correlation coefficient* provides a

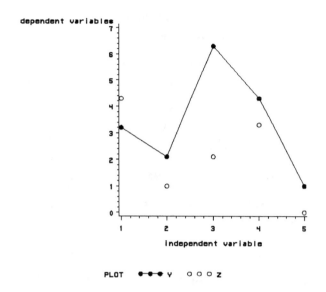

Figure 2.2: Overlaid scatterplots produced by **proc gplot**.

numerical summarization of the degree to which a linear relationship exists between two quantitative variables, and can be calculated using the **proc corr** command. For example, for the data of Example 2.4 in IPS, depicted in Figure 2.1, the commands

```
data archaeop;
input femur humerus;
cards;
38 41
56 63
59 70
64 72
74 84
proc corr data=archaeop;
var femur humerus;
run;
```

produce the output shown in Figure 2.3. It gives the mean, standard deviation, sum, minimum, and maximum of each variable and a 2×2 array that contains the value .99415 in the upper right and lower left corners. This is

```
                         Correlation Analysis

                2 'VAR' Variables:  FEMUR    HUMERUS


                         Simple Statistics

   Variable       N        Mean      Std Dev        Sum      Minimum      Maximum

   FEMUR          5     58.200000   13.198485   291.000000   38.000000   74.000000
   HUMERUS        5     66.000000   15.890249   330.000000   41.000000   84.000000


        Pearson Correlation Coefficients / Prob > |R| under Ho: Rho=0 / N = 5

                                        FEMUR          HUMERUS

                   FEMUR              1.00000          0.99415
                                      0.0              0.0005

                   HUMERUS            0.99415          1.00000
                                      0.0005           0.0
```

Figure 2.3: Correlation coefficient for Example 2.4 in IPS computed using **proc corr**.

the value of the correlation coefficient (sometimes called the Pearson correlation coefficient). The numbers below the correlation coefficient are P-values that are used to test whether or not the correlations are in fact 0. At this point in the course we ignore these numbers.

The following statements can be used in the **proc corr** procedure.

proc corr *options*;
var *variables*;
with *variables*;
weight *variable*;
freq *variable*;
by *variables*;

Some of the options that may appear in the **proc corr** statement are

data = *SASdataset*
outp = *name*
nomiss

To illustrate the use of these statements and options the data set **one** below contains the variables w, x, y, and z, and the commands

```
data one;
input w x y z;
```

```
Pearson Correlation Coefficients / Prob > |R| under Ho: Rho=0 / N = 5

                      X                 Y                 Z

    X             1.00000           0.80096           0.52568
                  0.0               0.1034            0.3629

    Y             0.80096           1.00000           0.51152
                  0.1034            0.0               0.3783

    Z             0.52568           0.51152           1.00000
                  0.3629            0.3783            0.0
```

Figure 2.4: Correlation matrix produced by **proc corr**.

```
cards;
.3 3.2 4.2 5.3
.2 1.1 2.0 1.0
.2 0.4 1.5 4.3
.1 1.2 1.4 3.2
.2 3.0 2.3 4.3
proc corr data=one;
var x y z;
run;
```

compute the correlations between variables x and y, x, and z, and y, and z. Part of the output is provided in Figure 2.4 (we have deleted the means, standard deviations and so on) in the form of a 3×3 *correlation matrix*. The correlation between x and y is .80096, between x and z it is .52568, and between y and z the correlation is .51152.

If the **var** statement is omitted, then correlations are computed between all numeric variables in the data set. If you want to produce correlations only for specific combinations, then use the **with** statement. For example,

```
proc corr data=one;
var x;
with y, z;
```

produces correlations between x and y, and x and z only.

To compute a *weighted correlation coefficient* use the **weight** statement. This is used when some observations are felt to be more important than others. If the i-th observation is given weight $w_i >= 0$, then the weighted correlation coefficient between $x_1, \ldots, x_n$ and $y_1, \ldots, y_n$ is

$$r_w = \frac{\sum_{i=1}^{n}[w_i(x_i - \bar{x}_w)(y_i - \bar{y}_w)]}{[\sum_{i=1}^{n} w_i(x_i - \bar{x}_w)^2 \sum_{i=1}^{n} w_i(y_i - \bar{y}_w)^2]^{\frac{1}{2}}}$$

where

$$\bar{x}_w = \frac{\sum_{i=1}^{n} w_i x_i}{\sum_{i=1}^{n} w_i}$$

and

$$\bar{y}_w = \frac{\sum_{i=1}^{n} w_i y_i}{\sum_{i=1}^{n} w_i}$$

are the weighted means. When $w_1 = \cdots = w_n = 1$, we get the usual correlation coefficient. Otherwise, observations with more weight have a larger influence on the computed value ($w_i = 0$ means no influence). If, in data set one, the variable w contains the weights, then

```
proc corr data=one;
var x y z;
weight w;
```

computes this correlation coefficient between x and y, x and z, and y and z. The weighted correlation between x and y is 0.82119.

If the i-th observation represents f_i observations – the pair (x_i, y_i) is observed f_i times – then the appropriate formula for the correlation is

$$r = \frac{\sum_{i=1}^{n} f_i(x_i - \bar{x}_f)(y_i - \bar{y}_f)}{[\sum_{i=1}^{n} f_i(x_i - \bar{x}_f)^2 \sum_{i=1}^{n} f_i(y_i - \bar{y}_f)^2]^{\frac{1}{2}}}$$

where

$$\bar{x}_f = \frac{\sum_{i=1}^{n} f_i x_i}{\sum_{i=1}^{n} f_i}$$

and

$$\bar{y}_f = \frac{\sum_{i=1}^{n} f_i x_i}{\sum_{i=1}^{n} f_i}$$

This formula agrees with the weighted case on taking $w_i = f_i$. To compute the correlation coefficients in this case, however, we must use the **freq** statement, as the test that a correlation coefficient is 0 that SAS does is different when the w_i are truly only weights and not counts.

If the **outp** = *name* option is specified, then SAS creates an output data set called *name* of a somewhat different structure. It is called a **type = corr** data set and it contains basic statistics and the Pearson correlation matrix for the variables in the **var** statement. Other SAS procedures recognize and use this type of data set.

The **nomiss** option specifies that any observations that have any missing values in them must be excluded from any calculations.

2.1.3 PROC REG

Regression is a technique for assessing the strength of a linear relationship between two variables. For regression we use **proc reg** command. Actually regression analysis applies to the analysis of many more variables than just two. We discuss more fully the **proc reg** procedure in Chapters II.10 and II.11.

As noted in IPS the regression analysis of two quantitative variables involves computing the least-squares line $y = a+bx$, where one variable is taken to be the response variable y and the other is taken to be the explanatory or predictor variable x. Note that the least-squares line is different depending upon which choice is made. For example, for the data of Example 2.4 in IPS and plotted in Figure 2.1, letting femur length be the response and humerus length be the explanatory variable, the commands

```
proc reg data = archaeop;
model femur = humerus;
run;
```

give the output in Figure 2.5. Much of this can be ignored at this point in the course. The table labeled `Parameter Estimates` gives the least-squares line as $y = 3.700990 + 0.825743x$, i.e., $a = 3.700990$ and $b = 0.825743$. Also the value of the square of the correlation coefficient is given as `R-Square`, which here equals .9883 or 98.83.We discuss the remaining output from the **regress** command in Chapter II.10.

```
Model: MODEL1
Dependent Variable: FEMUR

                        Analysis of Variance

                          Sum of        Mean
     Source      DF      Squares       Square    F Value    Prob>F

     Model        1     688.66931    688.66931    254.100    0.0005
     Error        3       8.13069      2.71023
     C Total      4     696.80000

          Root MSE      1.64628    R-square    0.9883
          Dep Mean     58.20000    Adj R-sq    0.9844
          C.V.          2.82866

                        Parameter Estimates

                    Parameter    Standard    T for H0:
     Variable  DF    Estimate       Error    Parameter=0    Prob > |T|

     INTERCEP   1    3.700990    3.49727376       1.058       0.3676
     HUMERUS    1    0.825743    0.05180152      15.941       0.0005
```

Figure 2.5: Output from application of **proc reg** to the data of Example 2.4 in IPS.

The following statements can appear with this procedure.

proc reg *options*;
model *dependent=independent /options*;
by *variables*;
freq *variable*;
id variable;
var *variables*;
weight *variable*;
plot *yvariable*xvariable = 'symbol' /options*;
output out = *SAS-dataset* **keyword** = *names*;

The following options are some of those that may appear in the **proc reg** statement.

data = *SASdataset*
corr
simple
noprint

The **model** statement causes the least-squares line of the form *dependent* = $a + b$ (*independent*) to be calculated, where *dependent* is the response variable and *independent* is the explanatory or predictor variable. There can be several **model** statements in a **proc reg**. The **by** statement works as described in other procedures; see **proc sort** for discussion. The **freq** statement identifies a variable that gives a count for the number of times that observation occurs in the original data set. The **id** statement identifies a variable whose values are used as an identifier for observations. This is useful in some of the diagnostic procedures where we want to identify influential or aberrant observations. The **var** statement must be used when several **model** statements are used, or only those variables that appear in the first **model** statement are available for subsequent analysis. Thus we must list all the variables we are going to use in the **var** statement. The **weight** statement identifies *variable* as containing weights for the dependent variable values.

The **plot** statement causes scatterplots to be produced. For example, the statements

```
proc reg data = archaeop;
model femur = humerus;
plot femur*humerus='0' residual.*humerus='+'
        residual.*predicted.='*'/ hplots =2 vplots=3;
run;
```

cause three scatterplots to be produced: `femur` versus `humerus`, `residual` versus `humerus`, and `residual` versus `predicted` values. Figure 2.6 is the plot of `residual` versus `humerus` that results from this command. Note the use of the period after the keyword names **predicted** and **residual** in this statement.

We may wish to compute other functions of residuals – predicted values, for example – or form other plots. Hence it is useful to be able to save these quantities in a SAS data set. This is accomplished using the **output** statement. For example,

```
proc reg data = archaeop noprint;
model femur = humerus;
output out=save predicted=yhat residual=r;
proc print data=save;
run;
```

creates a SAS data set called `save` with five observations and four variables `femur`, `humerus`, `yhat`, and `r`, where `yhat` is the predicted value for an observation and `r` is the residual for an observation, i.e., the difference between the observed value of `femur` and the predicted value. The program also prints

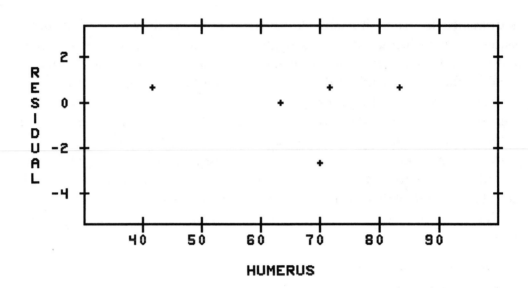

Figure 2.6: Residual plot for Example 2.4 in IPS produced by **proc reg**.

this data set and we show this output in Figure 2.7. In general all the variables in the original data set are included plus those defined. The format is to specify the statistic and name for the variable that will contain its values in the new data set via *statistic=name*. There are other statistics besides the predicted values and the residuals that we can save; we discuss them in Chapter II.10. If we want to create a data set with only some of these values, then the option **noprint** in the **proc reg** statement can be given to suppress printing.

Various options are available in the **proc reg** statement. Many **model** statements may appear in the procedure, and if an option appears in the **proc reg** statement it applies to all of them. For example, **corr** requests that the correlation matrix of all the variables in a **model** statement be printed, **simple** requests that the sum, mean, variance, standard deviation and uncorrected sum of squares be printed for each variable.

Several options can be used with the model statement.

noint causes the model $y = bx$ to be fit; i.e. no intercept term a is included.
p causes predicted values and ordinary residuals (difference between observed and predicted values) to be printed.

A useful plot for this context is obtained using **proc gplot**. The com-

OBS	FEMUR	HUMERUS	YHAT	R
1	38	41	37.5564	0.44356
2	56	63	55.7228	0.27723
3	59	70	61.5030	-2.50297
4	64	72	63.1545	0.84554
5	74	84	73.0634	0.93663

Figure 2.7: The saved date set **save** from **proc reg** applied to Example 2.4 in IPS.

mands

```
axis length=4 in;
symbol value=dot interpol=r;
proc gplot data=archaeop;
plot femur*humerus/ haxis=axis vaxis=axis;
```

produce a scatterplot of **femur** versus **humerus**, and the **symbol** statement with **interpol=r** causes the least-squares line to be plotted on this graph as well. This is shown in Figure 2.8.

2.2 Relationships Between Two Categorical Variables

The relationship between two categorical variables is typically assessed by crosstabulating the variables in a table. For this the **proc freq** and **proc tabulate** procedures are available.

We discussed **proc freq** in Section II.1.1.1, and we advise the reader to review that section. Here we simply add that **proc freq** has the capacity to cross tabulate variables as well as produce tabulations of single variables. For example,

```
data one;
input x y;
cards;
1 2
0 1
```

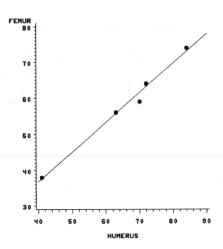

Figure 2.8: Scatterplot of data together with least-squares line in Example 2.4 of IPS produced using **proc gplot**.

```
2 2
0 2
2 2
1 1
2 1
2 1
1 2
proc freq data=one;
tables x*y;
run;
```

produces the 3×2 table given in Figure 2.9. To see if there is a relationship between the two variables we compare the conditional distributions of y given x, or the conditional distributions of x given y. In this case, comparing the conditional distributions of y given x, the three conditional distributions (.5,.5), (.33,.33) and (.5,.5) are different and so there would appear to be a relationship. Of course this is a small amount of data. In Chapters II.8 and II.9 we will see how to assess such a conclusion statistically.

If you want a *cross tabulation* table, then in the **tables** statement give the two variables for the table, separating the names with an asterisk *. Values of the first variable form the rows of the table, values of the second variable form

```
                          TABLE OF X BY Y

         X               Y

         Frequency|
         Percent  |
         Row Pct  |
         Col Pct  |          1|          2| Total
         ────────┼─────────┼─────────┤
              0  |        1|        1|       2
                 |    11.11|    11.11|   22.22
                 |    50.00|    50.00|
                 |    25.00|    20.00|
         ────────┼─────────┼─────────┤
              1  |        1|        2|       3
                 |    11.11|    22.22|   33.33
                 |    33.33|    66.67|
                 |    25.00|    40.00|
         ────────┼─────────┼─────────┤
              2  |        2|        2|       4
                 |    22.22|    22.22|   44.44
                 |    50.00|    50.00|
                 |    50.00|    40.00|
         ────────┼─────────┼─────────┤
         Total            4          5          9
                      44.44      55.56     100.00
```

Figure 2.9: Table resulting from cross tabulation using **proc freq**.

the columns. If you want a three-way (or n-way) cross tabulation table, join the three (or n) variables with asterisks. Values of the last variable form the columns, and values of the next-to-last variable form the rows. A separate table is produced for each level(or combination of levels) of the remaining variables. For example, the statements

```
proc freq data=example;
tables a*b*c;
```

produce m tables, where m is the number of different values for the variable a. Each table has the values of b down the side and the values of c across the top.

Variable lists can be used to specify many tables. For example,

```
tables (x1 - x3)*(y1 y2);
```

is equivalent to

```
tables x1*y1 x1*y2 x2*y1 x2*y2 x3*y1 x3*y2;
```

If the **page** option is included in the **proc freq** statement, then no more than one table is printed on a single page.

Often it is a good idea to graph the conditional distributions in bar charts to visually compare them. For example, using the SAS data set **one** we

Conditional distributions of y given x

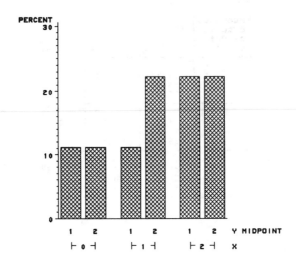

Figure 2.10: Side-by-side bar plots of the conditional distributions of y given x produced using **proc gplot**.

created, the commands

```
axis length = 4 in;
proc gchart data=one;
vbar y/type=pct group=x midpoints = 1 2
 maxis = axis raxis = axis;
title 'Conditional distributions of y given x';
```

create Figure 2.10, where the conditional distributions are plotted side-by-side using the **group** = *variable* option in the **vbar** statement.

2.3 Relationship Between a Categorical Variable and a Quantitative Variable

Suppose now that one variable is categorical and one is quantitative. We treat the situation where the categorical variable is explanatory and the quantitative variable is the response (the reverse situation is covered in Chapter II.15). To examine the relationship between such variables we look at the conditional distributions of the response variable given the explanatory variable. Since the response variable is quantitative, it is convenient to summarize these conditional distributions using means, standard deviations, or other summary statistics. To examine them in tabular form we use **proc tabulate**.

2.3.1 PROC TABULATE

This procedure constructs tables of descriptive statistics such as means, counts, standard deviations, and so on for cross-classified data. Each table cell contains a descriptive statistic calculated on all values of a response variable from observations sharing the same values of a set of categorical explanatory variable values. Note that this is different from **proc freq**, which only gives tables of counts or percentages for cross-classified data. Also the tables produced by **proc tabulate** can be more attractive because you have more control over their format.

To illustrate we use the data in Exercise 2.16 of IPS. Here we have four different colors of insect trap — lemon yellow, white, green and blue — and the number of insects trapped in six different instances in each trap. We have these data in a file called `c:\saslibrary\traps.dat` with variables `trap` and `count`. The variable `trap` takes the value 1 indicating a lemon yellow trap, 2 indicating white, 3 indicating green, and 4 indicating blue. The variable `count` is equal to the number of insects trapped in a particular trap. We then calculate the mean number of insects trapped for each trap using **proc tabulate**. The commands

```
data insect;
infile 'c:\saslibrary\traps.dat';
input trap count;
cards;
proc tabulate data=insect;
var count;
class trap;
table trap*count*mean ;
run;
```

produces a table (Figure 2.11) of mean counts for each of the four traps.

Following are some of the statements that can appear with **proc tabulate**.

proc tabulate *options*;

TRAP			
1	2	3	4
COUNT	COUNT	COUNT	COUNT
MEAN	MEAN	MEAN	MEAN
47.17	15.67	31.50	14.83

Figure 2.11: Table of means for the data of Exercise 2.16 in IPS produced by **proc tabulate**.

class *variables*;
var *variables*;
table *definition / options*;
freq *variable*;
weight *variable*;
by *variables*;
keylabel *keyword = text*;

An option available in the **proc tabulate** statement is

data = *SASdataset*

where *SASdataset* is a SAS data set containing the variables we want to tabulate.

The **proc tabulate** statement is always accompanied by one or more **table** statements specifying the tables to be produced. In the table statement, *definition* defines the **table** to be constructed and it can have a fairly elaborate structure. We discuss only cross tabulations here. We recognize two kinds of variables, namely, *class variables* and *analysis variables*. Class variables are identified in the **class** statement and analysis variables are identified in the **var** statement. Any of these variables can be crossed using the operator *. *Keywords* for statistics (such as **mean** and **std**) can also be crossed. When you cross class variables, categories are created from the combination of values of the variables. If one of the elements in a crossing is an analysis variable, then the statistics for the analysis variable are calculated for the categories created by the class variables.

The following keywords are used to specify statistics whose values will appear in the table cells. Only one of these can be specified in *definition*.

css corrected sum of squares.

cv coefficient of variation as a percentage.

max maximum value.

mean mean.

min minimum value.

n number of observations with nonmissing variable values.

nmiss number of observations with missing variable values.

range range = maximum - minimum

std standard deviation.

stderr standard error of the mean.

sum sum.

sumwgt sum of the weights.

uss uncorrected sum of squares

var variance.

Suppose that SAS data set **one** contains three categorical variables **a**, **b**, and **c** and one quantitative variable **x** and each of the categorical variables takes two values. Then the program

```
proc tabulate data=one;
var x;
class a b c;
table a * b * c * x * mean;
```

produces a table of means for **x** in each cell of the (a,b,c) classification in column format: the values of **a** form two columns, each of which is composed of two columns for **b**, each of which is composed of two columns for **c**, with the means of **x** written along the bottom of the table. The **table** statement

```
table a * b * x * mean;
```

produces a similar table, but now the means of **x** are for the (a,b) classification.

The **by**, **freq**, and **weight** statements occur once and apply to all tables defined in table statements. These statements work as described in **proc corr** and **proc means**. Many additional features of **proc tabulate** give you a great deal of control over the appearance of the tables. See reference 2 in Appendix E for more details.

2.4 Exercises

When the data for an exercise come from an exercise in IPS, the IPS exercise number is given in parentheses ( ). All computations in these exercises are to be carried out using SAS, and the exercises are designed to reinforce your understanding of the SAS material in this chapter. More generally, you should use SAS to do all the computations and plotting required for the problems in IPS.

1. (2.8) Calculate the least-squares line and make a scatterplot of Fuel used against Speed together with the least-squares line. Plot the residuals against Speed. What is the squared correlation coefficient between these variables?

2. (2.10) Make a scatterplot of Rate against Mass, labeling the points for males and females differently and including the least-squares line.

3. (2.17) Make a scatterplot of Weight against Pecking Order that includes the means and labels the points according to which pen they correspond to.

4. Create a SAS data set with 991 observations with two variables x and y, where x takes the values 1 through 100 with an increment of .1 and $y = x^2$. Calculate the correlation coefficient between x and y. Multiply each value in x by 10, add 5, and place the results in w. Calculate the correlation coefficient between y and w. Why are the correlation coefficients the same? Hint: To create the SAS data set use the method discussed in Section I.5.4.4.

5. Using the SAS data set created in Exercise 4, calculate the least-squares line with y as response and x as explanatory variable. Plot the residuals and describe the shape of the you observe. What transformation might you use to remedy the problem?

6. (2.40) For the data in this problem, numerically verify the algebraic relationship between the correlation coefficient and the slope of the least-squares line.

7. For Example 2.17 in IPS, calculate the least-squares line and reproduce Figure 2.21. Calculate the sum of the residuals and the sum of the

squared residuals. Divide the sum of the squared residuals by the number of data points minus 2. Is there anything you can say about what these quantities are equal to in general?

8. Suppose the observations in the following table are made on two categorical variables where variable 1 takes two values and variable 2 takes three values. Using **proc freq**, crosstabulate these data in a table of frequencies and in a table of relative frequencies. Calculate the conditional distributions of variable 1 given variable 2. Plot the conditional distributions in bar charts. Is there any indication of a relationship existing between the variables? How many conditional distributions of variable 2 given variable 1 are there?

Obs	Var 1	Var 2
1	0	2
2	0	1
3	0	0
4	1	0
5	1	2
6	0	1
7	1	2
8	0	0
9	0	1
10	1	1

9. Create a SAS data set consisting of two variables x and y where x takes the values 1 through 10 with an increment of .1 and $y = \exp(-1 + 2x)$. Calculate the least-squares line using y as the response variable and plot the residuals against x. What transformation would you use to remedy this residual plot? What is the least-squares line when you carry out this transformation?

10. (2.90) For the table given in this problem, use SAS commands to calculate the marginal distributions and the conditional distributions given field of study. Plot the conditional distributions.

Chapter 3

Producing Data

SAS statement introduced in this chapter

proc plan

This chapter is concerned with the collection of data, perhaps the most important step in a statistical problem because it determines the quality of whatever conclusions are subsequently drawn. A poor analysis can be fixed if the data collected are good simply by redoing the analysis. But if the data have not been appropriately collected, no amount of analysis can rescue the study. We discuss SAS statements and procedures that enable you to generate samples from populations and also to randomly allocate treatments to experimental units.

Once data have been collected, they are analyzed using a variety of statistical techniques. Virtually all of them involve computing *statistics* that measure some aspect of the data concerning questions we wish to answer. The answers determined by these statistics are subject to the uncertainty caused by the fact that we typically have not the full population but only a sample from the population. We therefore have to be concerned with the variability in the answers when different samples are obtained. This leads to a concern with the *sampling distribution* of a statistic. To assess the sampling distribution of a statistic, we make use of a powerful computational tool known as *simulation,* which we discuss in this and the following chapter.

SAS uses computer algorithms to mimic randomness. So the results are not truly random and in fact any simulation in SAS can be repeated, obtaining exactly the same results provided we start our simulation with the same *seed*.

3.1   PROC PLAN

Suppose we have a large population of size N and we want to select a sample of $n < N$ from the population. Further, suppose the elements of the population are ordered: a unique number $1, \ldots, N$ has been assigned to each element of the population. To avoid selection biases we want this to be a *random sample*; i.e., every subset of size n from the population has the same "chance" of being selected. We could do this physically using a simple random system such as chips in a bowl or coin tossing; we could use a table of random numbers, or, more conveniently, we can use computer algorithms that mimic the behavior of random systems.

For example, suppose there are 1000 elements in a population and we want to generate a sample of 100 from this population without replacement. We can use **proc plan** to do this. For example, the commands

```
proc plan seed=20398;
factors a=100 of 1000;
run;
```

generate the simple random sample of 100 from the set $\{1, 2, \ldots, 1000\}$; i.e., the commands generate a random sample from this set without replacement (Figure 3.1). If we were to run this procedure again with the same value for **seed**, we would get exactly the same sample. If you are going to generate multiple samples, be sure to change **seed** with each application of **proc plan** to ensure different samples. Note that **seed** must be any nonnegative integer less than or equal to $2^{31} - 1$.

Sometimes we want to generate *random permutations*, i.e., $m = n$ and we are simply reordering the elements of the population. For example, in experimental design suppose we have $n = n_1 + \cdots + n_k$ experimental units and k treatments. We want to allocate n_i applications of treatment i, and further, we want all possible such applications to be equally likely. Then we generate a random permutation $(l_1, \ldots, l_N)$ of $(1, \ldots, N)$ and allocate treatment 1 to experimental units labeled $l_1, \ldots, l_{n_1}$, allocate treatment 2 to experimental units labeled $l_{n_1+1}, \ldots, l_{n_1+n_2}$, and so on. The procedure **proc**

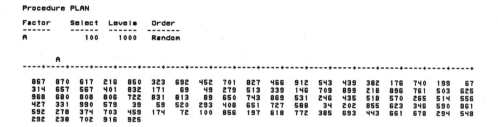

Figure 3.1: Simple random sample (sampling without replacement) of 100 from the numbers 1 through 1000 generated using **proc plan**.

plan can be used for this as well. For example,

```
proc plan seed=4449994;
factors a=25 of 25/noprint;
output out=one;
proc print data=one;
run;
```

generates a random permutation of $(1, 2, \ldots, 25)$ and outputs this to SAS data set **one**, as 25 observations of the variable **a**. The **noprint** option to the **factors** statement ensures that no output is printed in the Output window.

These examples show how to directly generate a sample from a population of modest size, but what happens if the population is huge or it is not convenient to label each unit with a number? For example, suppose we have a population of size 100,000 for which we have an ordered list and we want a sample of size 100. More sophisticated techniques need to be used, but simple random sampling can still typically be accomplished; see Exercise 3 for a simple method that works in some contexts.

3.2 Sampling from Distributions

Once we have generated a sample from a population, we measure various attributes of the sampled elements. For example, if we were sampling from a population of humans we might measure each sampled unit's height. The height for the sample unit is now a random quantity that follows the height distribution in the population we are sampling from. For example, if 80% of the people in the population are between 4.5 feet and 6 feet, then under *repeated sampling* of an element from the population (with replacement), in the long run the heights of 80% of the sampled units will be in this range.

Sometimes we want to sample directly from this population distribution, i.e., generate a number in such a way that under repeated sampling the proportion of values falling in any range agrees with that prescribed by the population distribution. Of course we typically don't know the population distribution; it is what we want to identify in a statistical investigation. Still there are many instances where we want to pretend that we do know it and simulate from this distribution. For perhaps we want to consider the effect of various choices of population distribution on the sampling distribution of some statistic of interest.

Computer algorithms allow us to generate random samples from a variety of different distributions. In SAS this is accomplished in the data step using the random number functions discussed in Appendix A.2.7. For example, suppose we want to simulate the tossing of a coin. Then the commands

```
data sample;
seed=1234556;
sum=0;
do i=1 to 100;
x=ranbin(seed,1,.75);
output sample;
sum=sum+x;
end;
prop=sum/100;
put prop;
drop seed i sum;
cards;
run;
```

generate a sample of 100 from the Bernoulli(.75) distribution. This sample is output to the SAS data set **sample** as the variable x. Also the program computes the proportion of 1's in the sample and outputs this value in the Log window using the **put** statement. For this run we obtained the value prop=.84. We used the **drop** statement to stop any variables we aren't interested in from being written to **sample**. The **drop** statement can appear anywhere in the data step. When we ran the program generating a sample of size 10^4 we got the value prop=.7472.

Often a normal distribution with some particular mean and standard deviation is considered a reasonable assumption for the distribution of a measurement in a population. For example,

```
data;
seed=67536;
sum=0;
do i=1 to 10000;
x=2+5*rannor(seed);
if x le 3 then sum =sum+1;
end;
prop=sum/10000;
put prop;
cards;
run;
```

generates a sample of 10^4 from the $N(2, 5)$ distribution, computes the proportion less than or equal to 3, and writes it in the Log window. In this case we got `prop=.5777`. The theoretically correct proportion is .5793.

3.3   Simulating Sampling Distributions

Once a sample is obtained, we compute various statistics based on these data. For example, suppose we flip a possibly biased coin n times and then want to estimate the unknown probability p of getting head. The natural estimate is $\hat{p}$ the proportion of heads in the sample. We would like to assess the sampling behavior of this statistic in a simulation. To do this we choose a value for p, then generate N samples from the Bernoulli distribution of size n, for each of these compute $\hat{p}$, then look at the empirical distribution of these N values, perhaps plotting a histogram as well. The larger N is, the closer the empirical distribution and histogram will be to the true sampling distribution of $\hat{p}$.

Note that there are two sample sizes here: the sample size n of the original sample the statistic is based on, which is fixed, and the *simulation* sample size N, which we can control. This is characteristic of all simulations. Sometimes, using more advanced analytical techniques, we can determine N so that the sampling distribution of the statistic is estimated with some prescribed accuracy. Some techniques for doing this are discussed in later chapters of

IPS. Another method is to repeat the simulation a number of times, slowly increasing N until we see the results stabilize. This is sometimes the only way available, but caution should be shown as it is easy for simulation results to be very misleading if the final N is too small.

We illustrate a simulation to determine the sampling distribution of $\hat{p}$ when sampling from a *Bernoulli*(.75) distribution. The commands

```
data dist;
seed= 345234;
do i=1 to 10000;
sum=0;
do j=1 to 25;
x=ranbin(seed,1,.2);
sum=sum+x;
end;
prop=sum/25;
drop i j sum seed x;
output dist;
end;
cards;
axis length = 4 in;
proc gchart data=dist ;
vbar prop / type=pct raxis=axis maxis=axis;
run;
```

generate 10^4 samples of 25 from the *Bernoulli*(.2) distribution, computes the proportion of 1's in this sample in the variable prop, outputs prop to the SAS data set dist, and then plots these 10^4 values of prop in a histogram. Note that we used the **drop** statement to eliminate any variables we weren't interested in having in the output data set.

3.4   Exercises

When the data for an exercise come from an exercise in IPS, the IPS exercise number is given in parentheses ( ). All computations in these exercises are to be carried out using SAS, and the exercises are designed to reinforce your understanding of the SAS material in this chapter. More generally, you

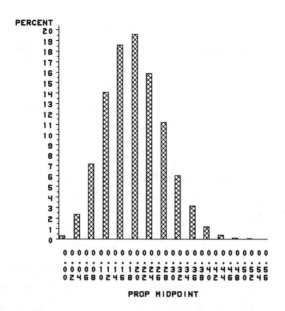

Figure 3.2: Histogram of 10^5 proportions of 1's generated in a sample of 25 from the *Bernoulli*(.2) distribution.

should use SAS to do all the computations and plotting required for the problems in IPS.

1. (3.14) Generate a random permutation of the names.

2. (3.27) Use the **proc sort** command to order the subjects by weight. Create five blocks of equal size by placing the four heaviest in the first block, the next four heaviest in the next block, and so on. Generate a random permutation of each block.

3. Use the following methodology to generate a sample of 20 from a population of 100,000. Repeatedly generate sequences of six uniformly distributed values in $\{0, 1, \ldots, 9\}$ until you obtain 20 unique sequences corresponding to numbers in $\{0, \ldots, 99,999\}$, and then select the 20 individuals in the population with these labels. Why does this work?

4. Suppose you want to carry out stratified sampling where there are three strata, the first stratum containing 500 elements, the second stratum

containing 400 elements, and the third stratum containing 100 elements. Generate a stratified sample with 50 elements from the first stratum, 40 elements from the second stratum and 10 elements from the third stratum. When the strata sample sizes are the same proportion of the total sample size as the strata population sizes are of the total population size this is called *proportional sampling*.

5. Carry out a simulation study with $N = 1000$ of the sampling distribution of $\hat{p}$ for $n = 5, 10, 20$ and for $p = .5, .75, .95$. In particular calculate the empirical distribution functions and plot the histograms. Comment on your findings.

6. Carry out a simulation study with $N = 2000$ of the sampling distribution of the sample standard deviation when sampling from the $N(0, 1)$ distribution based on a sample of size $n = 5$. In particular plot the histogram using midpoints 0, 1.5, 2.0, 2.5, 3.0, 5.0. Repeat this task for the sample coefficient of variation (sample standard deviation divided by the sample mean) using the midpoints -10, -9, ..., 0, ..., 9, 10. Comment on the shapes of the histograms relative to a $N(0, 1)$ density curve.

7. Suppose we have an urn containing 100 balls with 20 labeled 1, 50 labeled 2, and 30 labeled 3. Using sampling with replacement, generate a sample of size 1000 from this distribution. Hint: Use the random number function **rantbl**. Use **proc freq** to record the proportion of each label in the sample.

Chapter 4

Probability: The Study of Randomness

In this chapter of IPS the concept of probability is introduced more formally. Probability theory underlies a powerful computational methodology known as simulation. Simulation has many applications in probability and statistics and also in many other fields, such as engineering, chemistry, physics, and economics. We discussed some aspects of simulation in Chapter 3 and we continue here. We show how to do basic calculations and simulations in the data step. Actually, this is perhaps not the best way to do these kinds of calculations in SAS, as the data step is not designed for this purpose. For relatively small numbers of calculations this is not an issue, but if you are considering many calculations it would be better to use **proc iml** described in Appendix C.

4.1   Basic Probability Calculations

The calculation of probabilities for random variables can often be simplified by tabulating the cumulative distribution function. Also means and variances are easily calculated using SAS. For example, suppose we have the probability distribution

x	1	2	3	4
probability	.1	.2	.3	.4

in columns C1 and C2 with the values in C1 and the probabilities in C2. Then the commands

```
data calcul;
retain cum 0;
retain mean 0;
retain sum2 0;
input x p;
y1=x*p;
y2=x*x*p;
cum=cum +p;
mean=mean+y1;
sum2=sum2+y2;
var=sum2-mean**2;
if _N_=4 then
put 'mean = ' mean 'variance = ' var;
cards;
1 .1
2 .2
3 .3
4 .4
proc print data=calcul;
var x cum;
run;
```

input the observations, calculate the cumulative distribution function, the mean, and the variance of this distribution, and print the cumulative distribution function in the Output window and the mean and variance in the Log window. We use the implicit do index variable _N_ so that we write out only the mean and variance when we have finished inputting the data. Recall that _N_ is a variable that is set to 1 when the first observation is input and then is incremented by 1 as each subsequent observation is input. The **retain** statement takes the form

retain *variable initial-value*

and it causes *variable* to retain its value from one iteration of the data step to the next and *variable* is set equal to *initial-value* in the first iteration of the data step. There are other ways to carry out these calculations using SAS that are somewhat simpler. In particular arrays, discussed in Appendix B, are helpful in this regard.

4.2   Simulation

We already discussed and illustrated in Chapter 3 the use of the random number functions available in SAS (listed in Appendix A.2.7). We now illustrate some common simulations that we encounter in applications of statistics and probability.

4.2.1   Simulation for Approximating Probabilities

Simulation can be used to approximate probabilities. For example, suppose we are asked to calculate

$$P(.1 \leq X_1 + X_2 \leq .3)$$

when X_1, X_2 are both independent and follow the uniform distribution on the interval $(0, 1)$. Then the commands

```
data;
seed=1111111;
sum=0;
do i=1 to 10000;
x1=ranuni(seed);
x2=ranuni(seed);
y=x1+x2;
if .1 lt y & y lt .3 then
sum=sum+1;
end;
prop=sum/10000;
sd=sqrt(prop*(1-prop)/10000);
put 'estimated probability = ' prop 'standard error = ' sd;
cards;
run;
```

generate two independent $U(0,1)$ random variables 10^4 times, compute the proportion of times their sum lies in the interval $(.1,.3)$ and produce the output

```
estimated probability = 0.0368 standard error = 0.0018827044
```

in the Log window. We will see later that a good measure of the accuracy of this estimated probability is the *standard error of the estimate,* which in this case is given by $\sqrt{\hat{p}(1-\hat{p})/N}$ where $\hat{p}$ is the estimated probability and N is the Monte Carlo sample size. As the simulation size N increases, the law of large numbers says that $\hat{p}$ converges to the true value of the probability.

4.2.2   Simulation for Approximating Means

The means of distributions can also be approximated using simulations in SAS. For example, suppose X_1, X_2 are both independent and follow the uniform distribution on the interval $(0,1)$, and suppose we want to calculate the mean of $Y = 1/(1+X_1+X_2)$. We can approximate this mean in a simulation. The code

```
data;
seed=5671111;
sum=0;
sum2=0;
do i=1 to 10000;
x1=ranuni(seed);
x2=ranuni(seed);
y=1/(1+x1+x2);
sum=sum+y;
sum2=sum2+y**2;
end;
mean=sum/10000;
var=(sum2-mean**2)/9999;
sd=sqrt(var/10000);
put 'estimated mean = ' mean 'standard error = ' sd;
cards;
run;
```

generates 10^4 independent pairs of uniforms (X_1, X_2) and for each of these computes Y. The average $\bar{Y}$ of these 10^4 values of Y is the estimate of the mean of Y, and

$$\sqrt{\frac{1}{N(N-1)} \left[\sum_{i=1}^{N} Y_i^2 - \bar{Y}^2 \right]}$$

for $N = 10^4$ is the standard error of this estimate, which we will see provides an assessment of the accuracy of the estimate $\bar{Y}$. Finally the program outputs

```
estimated mean = 0.5239339101 standard error = 0.005368738
```

in the Log window. As the simulation size N increases, the law of large numbers says that the approximation converges to the true value of the mean.

4.3   Exercises

When the data for an exercise come from an exercise in IPS, the IPS exercise number is given in parentheses ( ). All computations in these exercises are to be carried out using SAS, and the exercises are designed to reinforce your understanding of the SAS material in this chapter. More generally, you should use SAS to do all the computations and plotting required for the problems in IPS.

1. Suppose we have the probability distribution

x	1	2	3	4	5
probability	.15	.05	.33	.37	.10

   Using SAS verify that this is a probability distribution. Make a bar chart (probability histogram) of this distribution. Tabulate the cumulative distribution. Calculate the mean and variance of this distribution. Suppose that three independent outcomes (X_1, X_2, X_3) are generated from this distribution. Compute the probability that $1 < X_1 \leq 4, 2 \leq X_2$ and $3 < X_3 \leq 5$.

2. (4.26) Indicate how you would simulate the game of roulette using SAS. Based on a simulation of $N = 1000$, estimate the probability of getting red and a multiple of 3. Also record the standard error of the estimate.

3. A probability distribution is placed on the integers 1, 2, ..., 100, where the probability of integer i is c/i^2. Determine c so that this is a probability distribution. What is the mean value? What is the 90-th percentile? Generate a sample of 20 from the distribution.

4. The expression e^{-x} for $x > 0$ is the density curve for what is called the *Exponential* (1) distribution. Plot this density curve in the interval from 0 to 10 using an increment of .1. The random number function **ranexp** can be used to generate from this distribution. Generate a sample of 1000 from this distribution and estimate its mean. Approximate the probability that a value generated from this distribution is in the interval (1,2). The general *Exponential* (λ) has a density curve, for $x > 0$ given by $\lambda^{-1}e^{-x/\lambda}$, where $\lambda > 0$ is a fixed constant. If X is distributed *Exponential* (1) then it can be shown that $Y = \lambda X$ is distributed *Exponential* (λ). Repeat the simulation with $\lambda = 3$. Comment on the values of the estimated means.

5. Suppose you carry out a simulation to approximate the mean of a random variable X and you report the value 1.23 with a standard error of .025. If you are then asked to approximate the mean of $Y = 3 + 5X$, do you have to carry out another simulation? If not, what is your approximation and what is the standard error of this approximation?

6. (4.50) Simulate 5 rounds of the game Keno where you bet on 10 each time. Calculate your total winnings (losses!).

7. Suppose a random variable X follows a $N(3, 2.3)$ distribution. Subsequently conditions change and no values smaller than -1 or bigger than 9.5 can occur; i.e., the distribution is conditioned to the interval $(-1, 9.5)$. Generate a sample of 1000 from the truncated distribution and use the sample to approximate its mean.

8. Suppose X is a random variable and follows a $N(0, 1)$ distribution. Simulate $N = 1000$ values from the distribution of $Y = X^2$ and plot these values in a histogram with midpoints 0, .5, 1, 1.5, ..., 15. Approximate the mean of this distribution. Now generate Y directly from its distribution, which is known as a *Chisquare*(1) distribution. In general the *Chisquare*(k) distribution can be generated from using the **rangam** random number function; namely if X is distributed

$Gamma(\alpha, 1)$, then $Y = 2X$ is distributed $Chisquare(2\alpha)$. Plot the Y values in a histogram using the same midpoints. Comment on the two histograms.

9. If X_1 and X_2 are independent random variables with X_1 following a $Chisquare(k_1)$ distribution and X_2 following a $Chisquare(k_2)$ distribution, then it is known that $Y = X_1 + X_2$ follows a $Chisquare(k_1 + k_2)$ distribution. For $k_1 = 1$, $k_2 = 1$, verify this empirically by plotting histograms with midpoints 0, .5, 1, 1.5, ..., 15 based on simulations of size $N = 1000$.

10. If X_1 and X_2 are independent random variables with X_1 following a $N(0, 1)$ distribution and X_2 following a $Chisquare(k)$ distribution, then it is known that

$$Y = \frac{X_1}{\sqrt{X_2/k}}$$

follows a $Student(k)$ distribution. Use this to generate a sample of 10^5 from the $Student(3)$ distribution. Plot a histogram with midpoints $-10, -9, ..., 9, 10$.

11. If X_1 and X_2 are independent random variables with X_1 following a $Chisquare(k_1)$ distribution and X_2 following a $Chisquare(k_2)$ distribution, then it is known that

$$Y = \frac{X_1/k_1}{X_2/k_2}$$

follows a $F(k_1, k_2)$ distribution. Use this to generate a sample of 10^5 from the $F(1, 1)$ distribution. Plot a histogram with midpoints 0, .5, 1, 1.5, ..., 15 based on simulations of size $N = 1000$.

Chapter 5

From Probability to Inference

SAS statement introduced in this chapter

proc shewhart

In this chapter the subject of statistical inference is introduced. Whereas we may feel fairly confident that the variation in a system can be described by probability, it is typical that we don't know which probability distribution is appropriate. Statistical inference prescribes methods for using data derived from the contexts in question to choose appropriate probability distributions. For example, in a coin-tossing problem the $Bernoulli(p)$ distribution is appropriate when the tosses are independent, but what is an appropriate choice of p?

5.1 Binomial Distribution

Suppose $X_1, \ldots, X_n$ is a sample from the $Bernoulli(p)$ distribution; i.e., $X_1, \ldots, X_n$ are independent realizations where each X_i takes the value 1 or 0 with probabilities p and $1 - p$, respectively. Then the random variable $Y = X_1 + \cdots + X_n$ equals the number of 1's in the sample and follows, as discussed in IPS, a $Binomial(n, p)$ distribution. Therefore Y can take on any of the values $0, 1, \ldots, n$ with positive probability. In fact an exact formula

113

can be derived for these probabilities.

$$P(Y = k) = \binom{n}{k} p^k (1 - p)^{n-k}$$

is the probability that Y takes the value k for $0 \leq k \leq n$. When n and k are small, this formula can be used to evaluate this probability but it is almost always better to use software like SAS to do it; when these values are not small, it is necessary. Also we can use SAS to compute the $Binomial(n, p)$ cumulative probability distribution, i.e., the probability contents of intervals $(-\infty, x]$.

For example, the SAS program

```
data;
x=probbnml(.3,20,6)-probbnml(.3,20,5);
put x;
cards;
run;
```

computes the $Binomial(20, .3)$ probability at 6 by evaluating the distribution function of this distribution at 6 and subtracting from the value of the distribution function at 5. The answer 0.1916389828 is printed in the Log window. Note that the function **probbnml** is used to compute the distribution function of the binomial distribution. The general form of this function is

probbnml(p, n, m)

where this gives the $Binomial(n, p)$ distribution function evaluated at $m \in \{0, \ldots, n\}$.

Should we also want to simulate from the $Binomial(n, p)$ distribution, we use the **ranbin** function. For example,

```
data;
seed=1123456;
x=ranbin(seed,40,.8);
put x;
cards;
run;
```

generates a sample of five values from the $Binomial(40, .8)$ distribution and prints the values obtained

33
26
37
27
28

in the Log window.

5.2 Control Charts

Control charts are used to monitor a process to ensure that it is under statistical control. There is a wide variety of such charts, depending on the statistic used for the monitoring and the test used to detect when a process is out of control.

5.2.1 PROC SHEWHART

If you have SAS/QC as part of your version of SAS, **proc shewhart** can be used to plot control charts. For example, the commands

```
data control;
seed=342999;
do i=1 to 100;
x=5+2*rannor(seed);
sub=ceil(i/5);
output;
drop seed;
end;
cards;
axis1 length= 8 in;
axis2 length= 4 in;
proc shewhart data=control graphics ;
xchart x*sub/ mu0=5 sigma0=2 haxis=axis1 vaxis=axis2;
run;
```

create a data set called `control` with variables i, x, and sub in the data step. The variable x consists of a random sample of 100 from the $N(5,2)$ distribution, the variable i is an index for these observations; and the variable sub groups the observations successively into subgroups of size 5. The

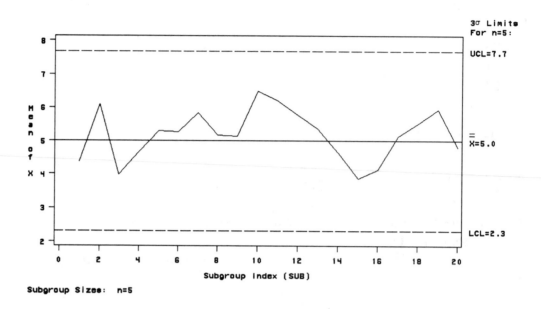

Figure 5.1: An $\bar{x}$ chart produced using **proc shewhart** with the theoretically correct values of μ and σ.

procedure is then used to create an $\bar{x}$ *chart* of these data, which is given in Figure 5.1. The 100 observations are partitioned into successive groups of 5 and $\bar{x}$ is plotted for each. The center line of the chart is at the mean 5 and the lines three standard deviations above and below the center line are drawn at $5 + 3 \cdot 2/\sqrt{5} = 7.683$ and at $5 - 3 \cdot 2/\sqrt{5} = 2.317$, respectively. The chart confirms that these choices for μ and σ are reasonable, as we might expect. The option **graphics** in the **proc shewhart** statement leads to a high-resolution plot in a Graph window otherwise a plot is provided in the Output window. Note that we had to create the grouping variable in the data step.

Of course we will typically not know the true values of μ and σ and these must be estimated from the data. The command

```
xchart x*sub/ haxis=axis1 vaxis=axis2;
```

draws the plot given in Figure 5.2. Here μ is estimated by the overall mean and σ is estimated by pooling the sample deviations for each subgroup. Again, this control chart indicates that everything is in order.

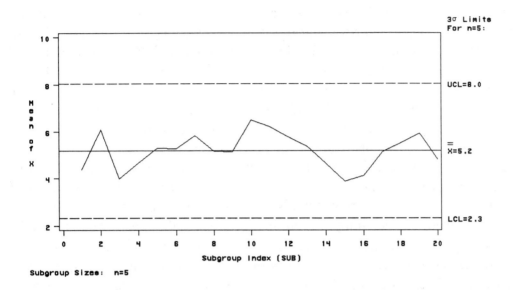

Figure 5.2: An $\bar{x}$ chart produced using **proc shewhart** with estimated values of μ and σ.

Following are some of the statements that can be used in **proc shewhart**.

proc shewhart *options*;
xchart *variable*grouping* / *options*;
pchart *variable*grouping* / *options*;
by *variables*;

When the **graphics** option is used, all the SAS/Graph statements we have discussed — axis, symbol and so on — are available for modifying the appearance of the plot.

The **xchart** statement indicates that an $\bar{x}$ chart is to be drawn with the means of the values of *variable* on the y axis for each subgroup and the values of *grouping* along the x axis. The following options can be used with the **xchart** statement.

mu0 $=$ *value*
sigma0 $=$ *value*
noconnect
alpha $=$ *value*
tests $=$ *values*

where the **mu0** specifies *value* as the mean of the population, **sigma0** specifies *value* as the standard deviation of the population, **noconnect** indicates
that the points in the plot should not be connected by lines, and **alpha**
specifies that the control limits should be such that a proportion of the observations, specified by *value,* should lie outside the control limits when the
process is under control. When alpha is not specified, the default is to draw
three sigma control limits. Using the **tests** option, various tests for control
can be carried out. For example, `tests = 1` checks to see if there is a least
one point outside the control limits, `tests =2` checks to see if there are nine
points in a row on one side of the central line, `tests = 3` checks to see if
there are six points in a row steadily increasing, etc. Eight different tests can
be performed, and as many as you like can be specified, e.g,. `tests = 2 3`.

The **pchart** statement works almost the same as **xchart** except that it
produces *p* charts. A *p* chart is appropriate when a response is coming from
a *Binomial* (n, p) distribution, e.g., the count of the number of defectives in
a batch of size n and we use the proportion of defectives $\hat{p}$ to control the
process. For example, the program

```
data control;
seed=342999;
do i=1 to 20;
x=ranbin(seed,55,.4);
output;
end;
cards;
axis1 length= 8 in;
axis2 length= 4 in;
proc shewhart data=control graphics ;
pchart x*i/ subgroupn = 55 haxis=axis1 vaxis=axis2 ;
run;
```

produces a plot like the high-resolution plot shown in Figure 5.3. Here 20 values are generated from a *Binomial*(55, .4) distribution as presumably arising
from a quality control process where 55 items were tested and the number
defective recorded each of the 20 times this was done. Therefore the proportion of defectives in each group of 55 is plotted against time as represented
by the index variable i, which is part of the data set `control` and presumably represents time here. The **pchart** statement has the same options as

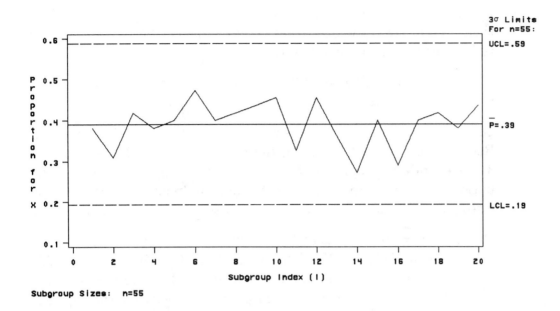

Figure 5.3: A p chart obtained using **proc shewhart**.

xchart with the exception that the **subgroupn** = *value* option must appear. It determines the number of items tested.

Many other control charts can be created using SAS/QC, and many other options can be used to control the plots. We refer the reader to references 8 and 9 in Appendix E.

5.3 Exercises

When the data for an exercise come from an exercise in IPS, the IPS exercise number is given in parentheses ( ). All computations in these exercises are to be carried out using SAS, and the exercises are designed to reinforce your understanding of the SAS material in this chapter. More generally, you should use SAS to do all the computations and plotting required for the problems in IPS.

1. Calculate all the probabilities for the *Binomial*(5, .4) distribution and the *Binomial*(5, .6) distribution. What relationship do you observe? Can you explain it and state a general rule?

2. Compute all the probabilities for a *Binomial*(5, .8) distribution and use them to directly calculate the mean and variance. Verify your answers using the formulas provided in IPS.

3. (5.17) Approximate the probability that in 50 polygraph tests given to truthful persons, at least one person will be accused of lying.

4. Generate $N = 1000$ samples of size $n = 5$ from the $N(0, 1)$ distribution. Record a histogram for $\bar{x}$ using the midpoints $-3, -2.5, -2, ..., 2.5,$ 3.0. Generate a sample of size $N = 1000$ from the $N(0, 1/\sqrt{5})$ distribution. Plot the histogram using the same midpoints and compare the histograms. What will happen to these histograms as we increase N?

5. Generate $N = 1000$ values of X_1, X_2, where X_1 follows a $N(3, 2)$ distribution and X_2 follows a $N(-1, 3)$ distribution. Compute $Y = X_1 - 2X_2$ for each of these pairs and plot a histogram for Y using the midpoints $-20, -15, ..., 25, 30$. Generate a sample of $N = 1000$ from the appropriate distribution of Y and plot a histogram using the same midpoints.

6. Plot the density curve for the *Exponential*(3) distribution (see Exercise II.4.4) between 0 and 15 with an increment of .1. Generate $N = 1000$ samples of size $n = 2$ from the *Exponential*(3) distribution and record the sample means. Standardize the sample of $\bar{x}$ using $\mu = 3$ and $\sigma = 3$. Plot a histogram of the standardized values using the midpoints $-5,$ $-4, ..., 4, 5$. Repeat this task for $n = 5, 10$. Comment on the shapes of the histograms. See Example 5.18 in IPS for further discussion of this distribution.

7. Plot the density of the uniform distribution on (0,1). Generate $N = 1000$ samples of size $n = 2$ from this distribution. Standardize the sample of $\bar{x}$ using $\mu = .5$ and $\sigma = \sqrt{1/12}$. Plot a histogram of the standardized values using the midpoints $-5, -4, ..., 4, 5$. Repeat this task for $n = 5, 10$. Comment on the shapes of the histograms.

8. The *Weibull* (β) has density curve $\beta x^{\beta-1} e^{-x^{\beta}}$ for $x > 0$, where $\beta > 0$ is a fixed constant. Plot the *Weibull* (2) density in the range 0 to 10 with an increment of .1. See Section 5.2 in IPS for discussion of this distribution.

9. (5.50) Make an $\bar{x}$ chart for these data with three sigma control lines using **proc shewhart**. What tests for control does the chart fail?

9. (5.50) Make an $\bar{x}$ chart for these data with three sigma control lines using **proc shewhart**. What tests for control does the chart fail?

10. (5.59) Make a p chart for these data with three sigma control lines using **proc shewhart**. What tests for control does the chart fail?

Chapter 6

Introduction to Inference

In this chapter the basic tools of statistical inference are discussed. There are a number of SAS commands that aid in the computation of confidence intervals and in carrying out tests of significance.

6.1 z Intervals and z tests

We want to make inference about the mean μ using a sample $x_1, \ldots, x_n$ from a distribution where we know the standard deviation σ. The methods of this section are appropriate in three situations.

(1) We are sampling from a normal distribution with unknown mean μ and known standard deviation σ and thus

$$z = \frac{\bar{x} - \mu}{\sigma/\sqrt{n}}$$

is distributed $N(0, 1)$.

(2) We have a large sample from a distribution with unknown mean μ and known standard deviation σ and the central limit theorem approximation to the distribution of $\bar{x}$ is appropriate, i.e.,

$$z = \frac{\bar{x} - \mu}{\sigma/\sqrt{n}}$$

is approximately distributed $N(0, 1)$.

(3) We have a large sample from a distribution with unknown mean μ and unknown standard deviation σ and the sample size is large enough so that

$$z = \frac{\bar{x} - \mu}{s/\sqrt{n}}$$

is approximately $N(0, 1)$, where s is the sample standard deviation.

The z confidence interval takes the form $\bar{x} \pm z^*\sigma/\sqrt{n}$, where s is substituted for σ in case (3) and z^* is determined from the $N(0, 1)$ distribution by the confidence level desired, as described in IPS. Of course situation (3) is probably the most realistic, but note that the confidence intervals constructed for (1) are exact while those constructed under (2) and (3) are only approximate and a larger sample size is required in (3) for the approximation to be reasonable than in (2).

Consider Example 6.2 in IPS and suppose the data 190.5, 189.0, 195.5, 187.0 are stored in the SAS data set **weights** in the variable **wt**. Using, as in the text, $\sigma = 3$ and $n = 4$, the program

```
proc means data = weights noprint;
var wt;
output out=calc mean=mnwt;
data;
set calc;
std=3/sqrt(4);
zl=mnwt-std*probit(.95);
zu=mnwt+std*probit(.95);
put '.90 confidence interval is (' zl ',' zu ')';
run;
```

calculates the mean of **wt** using **proc means** with no output to the Output window because of the **noprint** option, but it creates the SAS data set **calc**, which contains a single observation, and the variable mnwt, which contains the mean of **weight**. In the data step the 90% confidence interval (zl,zu) is computed using the inverse distribution function for the $N(0, 1)$ distribution via the **probit** function. The 90% confidence interval for μ

```
.90 confidence interval is (188.03271956 ,192.96728044 )
```

is written on the Log window using the **put** command.

Suppose we want to test the hypothesis that the unknown mean μ equals a value μ_0 and one of the situations (1), (2) or (3) obtains. The z test is based on computing a P-value using the observed value of

$$z = \frac{\bar{x} - \mu_0}{\sigma/\sqrt{n}}$$

and the $N(0,1)$ distribution as described in IPS.

Consider Example 6.6 in IPS, where we are asked to test the null hypothesis $H_0 : \mu = 187$ against the alternative $H_a : \mu > 187$. Suppose the data 190.5, 189.0, 195.5, 187.0 are stored in the SAS data set `weights` in the variable `wt`. Using, as in the text, $\sigma = 3$ and $n = 4$, the program

```
proc means data = weights noprint;
var wt;
output out=calc mean=mnwt;
data interval;
set calc;
std=3/sqrt(4);
z=(mnwt-187)/std;
p=1-probnorm(z);
put 'z = ' z 'P-value = ' p;
run;
```

calculates the z statistic and the P-value $P(Z > z) = 1 - P(Z \le z)$, where $Z \sim N(0,1)$. The values

```
z = 2.3333333333 P-value = 0.0098153286
```

are printed in the Log window. If we want to test $H_0 : \mu = 187$ against the alternative $H_a : \mu \ne 187$, then the relevant program is

```
proc means data = weights noprint;
var wt;
output out=calc mean=mnwt;
data interval;
set calc;
std=3/sqrt(4);
z=abs((mnwt-187)/std);
p=2*(1-probnorm(z));
put 'z = ' z 'P-value = ' p;
```

which computes $|z|$ using the **abs** function and the P-value $P(|Z| > |z|) = 2\left(1 - P(Z \le z)\right),$ where $Z \sim N(0,1)$. The values

```
z = 2.3333333333 P-value = 0.0196306573
```

are printed in the Log window.

6.2   Simulations for Confidence Intervals

When we are sampling from a $N(\mu, \sigma)$ distribution and know the value of σ, the confidence intervals constructed in II.6.1 are exact; in the long run a proportion 95% of the 95% confidence intervals constructed for an unknown mean μ will contain the true value of this quantity. Of course any given confidence interval may or may not contain the true value of μ, and in any finite number of such intervals so constructed, some proportion other than 95% will contain the true value of μ. As the number of intervals increases, however, the proportion covering will go to 95%.

We illustrate this via a simulation study based on computing 90% confidence intervals. The program

```
data conf;
seed=65398757;
z=probit(.95);
p=0;
do i=1 to 25;
x1=0;
do j=1 to 5;
x1= x1+(1+2*rannor(seed));
end;
x1=x1/5;
l=x1-z*2/sqrt(5);
u=x1+z*2/sqrt(5);
if l le 1 & 1 le u then
p=p+1;
output conf;
end;
p=p/25;
me=3*sqrt(p*(1-p)/25);
lp=p-me;
```

```
up=p+me;
put p '(' lp ',' up ')';
drop seed z j x1 p me lp up;
cards;
symbol1 value= dot color=black;
symbol2 value= circle color=black;
axis1 length= 8 in;
axis2 length= 4 in label = ('limits');
proc gplot data = conf;
plot l*i=1 u*i=2/overlay haxis=axis1 vaxis=axis2 vref=1;
run;
```

generates 25 samples of size 5 from the $N(1,2)$ distribution, calculates the lower endpoint of the 90% confidence interval in the variable l and the upper endpoint in the variable u, outputs these values to the data set conf, calculates the proportion of these confidence intervals that contain the true value of $\mu = 1$ in p together with the half-length of the interval that contains the true proportion with virtual certainty in me, and then outputs p together with this interval in the Log window. Finally, **proc gplot** is used to draw Figure 6.1, which plots each confidence interval together with a reference line at 1 (using the **vref** option). If the lower endpoint (black dot) is above this reference line, or the upper endpoint (open circle) is below this reference line, then the particular interval does not contain the true value. In this case the ouptut

```
0.88 (0.6850230783 ,1.0749769217 )
```

was recorded in the Log wndow, indicating that 88% of the intervals covered. The interval $(0.6850230783, 1.0749769217)$ indicates, however, that not much reliability can be placed in the estimate. When we repeated the simulation with 10^4 samples, we obtained

```
0.8981 (0.8890244972 ,0.9071755028 )
```

which indicates considerably greater accuracy. Note that in plotting Figure 6.1, we made use of the **label** = '*text*' option to the **axis** statement to label the vertical axis.

The simulation just carried out simply verifies a theoretical fact. On the other hand, when we are computing approximate confidence intervals — when we are not necessarily sampling from a normal distribution — it is good to do some simulations from various distributions to see how much

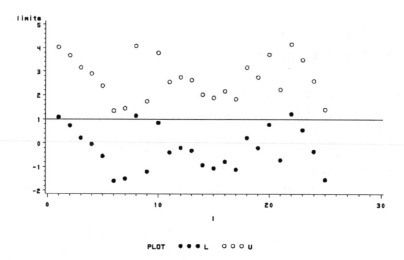

Figure 6.1: Plot of 25 .90 confidence intervals for the mean of a normal distribution where 25 samples of size 5 were generated from an $N(1,2)$ distribution.

reliance we can place in the approximation at a given sample size. The true *coverage probability* of the interval — the long-run proportion of times that the interval covers the true mean — will not in general be equal to the nominal confidence level. Small deviations are not serious, but large ones are.

6.3   Simulations for Power Calculations

It is useful to know in a given context how sensitive a particular test of significance is. By this we mean how likely it is that the test will lead us to reject the null hypothesis when the null hypothesis is false. This is measured by the concept of the *power* of a test. Typically a level α is chosen for the P-value at which we would definitely reject the null hypothesis if the P-value is smaller than α. For example, $\alpha = .05$ is a common choice for this level. Suppose then that we have chosen the level of .05 for the two-sided z test and we want to evaluate the power of the test when the true value of the mean is $\mu = \mu_1$; i.e., we want to evaluate the probability of getting a P-value smaller than .05 when the mean is μ_1. The two-sided z test with level α

rejects $H_0 : \mu = \mu_0$ whenever

$$P\left(|Z| > \left|\frac{\bar{x} - \mu_0}{\sigma/\sqrt{n}}\right|\right) \leq \alpha$$

where Z is a $N(0, 1)$ random variable. This is equivalent to saying that the null hypothesis is rejected whenever

$$\left|\frac{\bar{x} - \mu_0}{\sigma/\sqrt{n}}\right|$$

is greater than or equal to the $1 - \alpha/2$ percentile for the $N(0, 1)$ distribution. For example, if $\alpha = .05$, then $1 - \alpha/2 = .975$, and this percentile can be obtained via the **probit** function, which gives the value 1.96. Denote this percentile by z^*. Now if $\mu = \mu_1$ then

$$\frac{\bar{x} - \mu_0}{\sigma/\sqrt{n}}$$

is a realized value from the distribution of $Y = \frac{\bar{X} - \mu_0}{\sigma/\sqrt{n}}$ when $\bar{X}$ is distributed $N(\mu_1, \sigma/\sqrt{n})$. Therefore Y follows a $N(\frac{\mu_1 - \mu_0}{\sigma/\sqrt{n}}, 1)$ distribution. Then the power of the two-sided test at $\mu = \mu_1$ is

$$P(|Y| > z^*)$$

and this can be evaluated exactly using the **probnorm** function after writing

$$
\begin{aligned}
P(|Y| > z^*) &= P(Y > z^*) + P(Y < -z^*) \\
&= P\left(Z > -\frac{(\mu_1 - \mu_0)}{\sigma/\sqrt{n}} + z^*\right) + P\left(Z < -\frac{(\mu_1 - \mu_0)}{\sigma/\sqrt{n}} - z^*\right)
\end{aligned}
$$

with Z following a $N(0, 1)$ distribution.

This derivation of the power of the two-sided test depends on the sample coming from a normal distribution so that $\bar{X}$ has an exact normal distribution. In general, however, $\bar{X}$ will only be approximately normal, so the normal calculation is not exact. To assess the effect of the non-normality, however, we can often simulate sampling from a variety of distributions and estimate the probability $P(|Y| > z^*)$. For example, suppose that we want to test $H_0 : \mu = 0$ in a two-sided z test based on a sample of 10, where we estimate σ by the sample standard deviation and we want to evaluate the power at 3. Let us further suppose that we are actually sampling from a uniform distribution on the interval $(-10, 16)$, which indeed has its mean at 3. Then the simulation

```
data;
seed= 83545454;
p=0;
do i=1 to 10000;
x=0;
s2=0;
do j=1 to 10;
z=-10+26*ranuni(seed);
x=x+z;
s2=s2+z*z;
end;
x=x/10;
s2=(s2-x*x)/9;
y=x/sqrt(s2/10);
if abs(y) ge 1.96 then
p=p+1;
end;
p=p/10000;
ep=sqrt(p*(1-p)/10000);
put 'Estimate of power =' p 'with standard error =' ep;
cards;
run;
```

generates 10^4 samples of size 10 from the $U(-10, 16)$ distribution and calculates the proportion of times the null hypothesis is rejected in the variable p. The output

```
Estimate of power =0.1342 with standard error =0.0034086707
```

is written in the Log window and gives the estimate of the power as .1342 and the standard error of this estimate as approximately .003. The application determines whether or not the assumption of a uniform distribution makes sense and whether or not this power is indicative of a sensitive test or not.

6.4 Chi-square Distribution

If Z is distributed according to the $N(0,1)$ distribution, then $Y = Z^2$ is distributed according to the $Chisquare(1)$ distribution. If X_1 is distributed $Chisquare(k_1)$ independent of X_2 distributed $Chisquare(k_2)$, then $Y = X_1 + X_2$ is distributed according to the $Chisquare(k_1 + k_2)$ distribution. SAS commands assist in carrying out computations for the $Chisquare(k)$ distribution. Note that k is any nonnegative value and is referred to as the *degrees of freedom*. The density curve of the $Chisquare(k)$ distribution is given by the formula

$$f(x) = \frac{1}{\Gamma\left(\frac{k}{2}\right)} \left(\frac{x}{2}\right)^{\frac{k}{2}-1} \exp\left\{-\frac{x}{2}\right\} \left(\frac{1}{2}\right)$$

for $x > 0$ and where $\Gamma(w)$ can be evaluated using the **gamma** function in SAS. For example, suppose we want to plot the $Chisquare(10)$ density function in the interval $(0, 50)$. Then the program

```
data density;
const=2*gamma(10/2);
do i=1 to 1000;
x=i*30/1000;
f=((x/2)**4)*exp(-x/2)/const;
output density;
drop i const;
end;
cards;
axis1 length=4 in;
axis2 length=6 in;
symbol1 interpol=join;
proc gplot data=density;
plot f*x=1/ vaxis=axis1 haxis=axis2;
run;
```

calculates this density in the variable **f** at 1000 equispaced values of the variable **x** between 0 and 30 and then plots the curve in a scatterplot of **f** against **x**, shown in Figure 6.2.

The **probchi** and **cinv** functions are used to calculate the cumulative distribution and inverse cumulative distribution functions of the chi-square distribution. For example, the statements

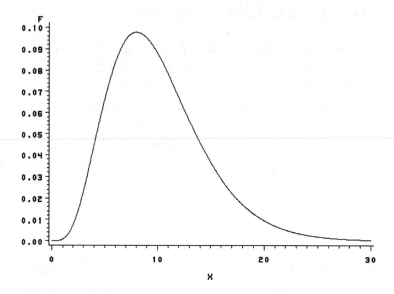

Figure 6.2: Plot of the *Chisquare*(10) density curve.

```
data;
x=probchi(3,5);
put 'x= ' x;
p=cinv(.68,7);
put 'p= ' p;
cards;
run;
```

calculate the value of the *Chisquare*(2) distribution function at 3 and the inverse *Chisquare*(7) distribution function at .68 — i.e., a .68-quantile — and write

```
x= 0.3000141641
p= 8.1447886689
```

in the Log window.

To generate samples from the chi-square distribution we use the **rangam** function. For example,

```
data;
seed=3241222;
```

```
do i=1 to 100;
x=2*rangam(seed,5);
put x;
end;
run;
```

generates a sample of 100 from the $Chisquare(10)$ distribution and prints these values in the Log window. Notice that we need to multiply

rangam(*seed,value*)

by 2 and that the degrees of freedom of the chi-square distribution being generated from equals 2(*value*).

Applications of the chi-square distribution are given later in the book, but we mention one here. In particular, if $x_1, \ldots, x_n$ is a sample from a $N(\mu, \sigma)$ distribution then $(n-1)\, s^2/\sigma^2 = \sum_{i=1}^{n} (x_i - \bar{x})^2 /\sigma^2$ is known to follow a $Chisquare(n-1)$, distribution and this fact is used as a basis for inference about σ (confidence intervals and tests of significance). Because of their nonrobustness to small deviations from normality, these inferences are not recommended.

6.5 Exercises

When the data for an exercise come from an exercise in IPS, the IPS exercise number is given in parentheses ( ). All computations in these exercises are to be carried out using SAS, and the exercises are designed to reinforce your understanding of the SAS material in this chapter. More generally, you should use SAS to do all the computations and plotting required for the problems in IPS.

1. (6.9) Use SAS to compute 90%, 95%, and 99% confidence intervals for μ.

2. (6.39) Use SAS to test the null hypothesis against the appropriate alternative. Evaluate the power of the test with level $\alpha = .05$ at $\mu = 33$.

3. Simulate $N = 1000$ samples of size 5 from the $N(1,2)$ distribution and calculate the proportion of .90 confidence intervals for the mean that cover the true value $\mu = 1$.

4. Simulate $N = 1000$ samples of size 10 from the uniform distribution on $(0,1)$ and calculate the proportion of .90 confidence intervals for the mean that cover the true value $\mu = .5$. Use $\sigma = 1/\sqrt{12}$.

5. Simulate $N = 1000$ samples of size 10 from the *Exponential*(1) distribution (see Exercise II.4.4) and calculate the proportion of .95 confidence intervals for the mean that cover the true value $\mu = 1$. Use $\sigma = 3$.

6. The density curve for the *Student*(1) distribution takes the form

$$\frac{1}{\pi} \frac{1}{1 + x^2}$$

for $-\infty < x < \infty$. This special case is called the *Cauchy* distribution. Plot this density curve in the range $(-20, 20)$ at 1000 equispaced points. Simulate $N = 1000$ samples of size 5 from the *Student*(1) distribution (see Exercise II.4.3.10) and using the sample standard deviation for σ, calculate the proportion of .90 confidence intervals for the mean that cover the value $\mu = 0$. It is possible to obtain very bad approximations in this example because the central limit theorem does not apply to the distribution. In fact it does not have a mean.

7. The uniform distribution on the interval (a, b) has mean $\mu = (a + b)/2$ and standard deviation $\sigma = \sqrt{(b - a)^2 / 12}$. Calculate the power at $\mu = 1$ of the two-sided z test at level $\alpha = .95$ for testing $H_0 : \mu = 0$ when the sample size is $n = 10$, σ is the standard deviation of a uniform distribution on $(-10, 12)$, and we are sampling from a normal distribution. Compare your result with the example in Section II.6.4.

8. Suppose we are testing $H_0 : \mu = 0$ in a two-sided test based on a sample of 3. Approximate the power of the z test at level $\alpha = .1$ at $\mu = 5$ when we are sampling from the distribution of $Y = 5 + W$ where W follows a *Student*(6) distribution (see Exercise II.4.3.10) and we use the sample standard deviation to estimate σ. Note that the mean of the distribution of Y is 5.

Chapter 7

Inference for Distributions

SAS statement introduced in this chapter

 proc ttest

7.1 Student Distribution

If Z is distributed $N(0,1)$ independent of X distributed $Chisquare(k)$ (see II.6.4) then

$$T = \frac{Z}{\sqrt{X/k}}$$

is distributed according to the $Student(k)$ distribution. The value k is referred to as the *degrees of freedom* of the Student distribution. SAS functions assist in carrying out computations for this distribution.

The density curve for the $Student(k)$ distribution can be plotted using the method of section II.6.4; see Exercise II.7.6.1. Also the functions **probt** and **tinv** can be used to obtain the values of the $Student(k)$ cumulative distribution function and the inverse distribution function, respectively. For example,

```
data;
p=probt(5,2);
x=tinv(.025,5);
put 'p= ' p 'x=' x;
run;
```

writes

```
p= 0.9811252243 x=-2.570581836
```

in the Log Window. To generate a value T from the $Student(k)$, generate a value $Z \sim N(0,1)$ and a value $X \sim Chisquare(k)$ and put $T = X/\sqrt{X/k}$.

7.2   The t Interval and t Test

When sampling from the $N(\mu, \sigma)$ distribution with μ and σ unknown, an exact $1 - \alpha$ confidence interval for μ based on the sample $x_1, \ldots, x_n$ is given by $\bar{x} \pm t^*s/\sqrt{n}$, where t^* is the $1 - \alpha/2$ percentile of the $Student(n-1)$ distribution. These intervals can be easily computed in SAS using the **tinv** function. Suppose the SAS data set **one** contains 20 observations on the variable **x**, which were generated from the $N(6,1)$ distribution, and we assume that the values of μ and σ are unknown. Then the program

```
proc means data=one noprint;
var x;
output out=calc mean=mnx std=stdx;
data interval;
set calc;
n=_freq_;
cl=mnx-(stdx/sqrt(n))*tinv(.975,n-1);
cu=mnx+(stdx/sqrt(n))*tinv(.975,n-1);
proc print data=interval;
var cl cu;
run;
```

calculates the mean and standard deviation of **x** in **proc means** and outputs these values to the data set **calc** as **mnx** and **stdx** respectively. The next data step creates a SAS data set interval by reading in **calc** and adding the variables **cl** and **cu**, which correspond to the lower and upper endpoints of a .95 confidence interval for μ. The SAS data set **calc** contains one observation and four variables **_type_**, **_freq_**, **mnx** and **stdx**. The variable **_freq_** is the sample size n in this case. Finally, **proc print** is used to print the confidence interval

OBS	CL	CU
1	4.08516	7.42346

in the Output window.

Suppose we have a sample $x_1, \ldots, x_n$ from a normal distribution, with unknown mean μ and standard deviation σ, and we want to test the hypothesis that the unknown mean equals a value μ_0. The test is based on computing a P-value using the observed value of

$$t = \frac{\bar{x} - \mu_0}{s/\sqrt{n}}$$

and the *Student*$(n-1)$ distribution as described in IPS. For example, if we want to test $H_0 : \mu = 6$ versus the alternative $H_a : \mu \neq 6$ for the variable x in the data set **one**, then the program

```
proc means data=one noprint;
var x;
output out=calc mean=mnx std=stdx;
data interval;
set calc;
n=_freq_;
t=abs(sqrt(n)*(mnx-6)/stdx);
pval=2*(1-probt(t,n-1));
proc print data=interval;
var t pval;
run;
```

calculates the t statistic in the variable t and the P-value for this two-sided test in the variable pval and prints the values

OBS	T	PVAL
1	0.30808	0.76137

in the Output window. Similarly, we can compute the P-values for the one-sided tests. The two-sided t test can also be carried out using **proc means**. For example, the statements

```
data two;
set one;
y=x-6;
proc means data=two t prt;
var y;
run;
```

result in the value of the t statistic together with the P-value for $H_0 : \mu = 6$ versus the alternative $H_a : \mu \neq 6$ being printed in the Output window.

We can also use this approach to construct the t intervals and carry out the t test in a *matched pairs design*. We create a variable equal to the difference of the measurements and apply the above analysis to this variable.

We can calculate the power of the t test using simulations. Note, however, that we must prescribe not only the mean μ_1 but the standard deviation σ_1 as well, as there are two unknown parameters. For example, the program

```
data;
seed=23734;
n=20;
mu0=6;
mu1=4;
sigma1=3;
p=0;
t0=tinv(.975,n-1);
do i=1 to 10000;
xbar=mu1+sigma1*rannor(seed)/sqrt(n);
x=(sigma1**2)*(2*rangam(seed,(n-1)/2))/(n-1);
t=abs(sqrt(n)*(xbar-mu0)/sqrt(x));
if t gt t0 then
p=p+1;
end;
p=p/10000;
stdp=sqrt(p*(1-p)/10000);
put p stdp;
run;
```

carries out a simulation to approximate the power of the t test for testing $H_0 : \mu = 6$ versus the alternative $H_a : \mu \neq 6$ at level $\alpha = .05$ with $\mu_1 = 4$, $\sigma_1 = 3$, and $n = 20$. The value of xbar is a randomly generated value of $\bar{x}$ from the $N(2,3)$ distribution, and the value of x is $\sigma_1^2 / (n-1)$ times a randomly generated value from the $Chisquare(19)$ distribution (note that $(n-1)s^2/\sigma_1^2 \sim Chisquare(n-1)$). The test statistic t is calculated in the variable t and the null hypothesis rejected whenever t>t0, where t0 is the .975 quantile of the $Student(n-1)$ distribution. In this case, we generated 10^4 values of t and recorded the proportion of rejections in p, which is the estimate of the power. This proportion together with standard error equal

0.8068 0.0039480851

which is written in the Log window. See Exercise 9 for more relevant discussion.

7.3 The Sign Test

As discussed in IPS, sometimes we cannot sensibly assume normality or transform to normality or make use of large samples so that there is a central limit theorem effect. In such a case we attempt to use *distribution-free* or *nonparametric* methods. The *sign test* for the median is one such method.

For example, suppose we have the data for Example 7.1 in IPS in the variable vitc in SAS data set ex71. Then the program

```
data test;
set ex71;
y=vitc-20;
proc univariate data=test noprint;
var y;
output out=results msign=sigstat probs=pvals;
proc print data=results;
var sigstat pvals;
run;
```

uses **proc univariate** to output to the SAS data set **results**. The data set **results** contains the value of the sign statistic in **sigstat** and the P-value in **pvals**, based on the variable y=vitc-20, for testing $H_0 : \zeta = 0$ versus $H_0 : \zeta \neq 0$ where ζ denotes the population median. The values

OBS	SIGSTAT	PVALS
1	2 0	.35938

are then printed in the Output window. In this case the P-value of .35938 indicates that we would not reject the null hypothesis that the median of the distribution of vitc is 20.

The test statistic calculated by **proc univariate** is M(Sign), which is the sign test statistic minus its mean — $n/2$ where n is the sample size — under $H_0 : \zeta = 0$ and the P-value for testing H_0 against $H_a : \zeta \neq 0$ is also computed. Denote this P-value by P_2. Suppose we want to instead test $H_0 : \zeta = 0$ versus $H_a : \zeta > 0$. Then if M(Sign)> 0, the relevant P-value is $.5P_2$, and if M(Sign)< 0, the relevant P-value is $1 - .5P_2$. Say we want to test $H_0 : \zeta = 0$ versus $H_a : \zeta < 0$. Then if M(Sign)< 0, the relevant P-value is $.5P_2$, and if M(Sign)> 0, the relevant P-value is $1 - .5P_2$. We can also use the sign test when we have paired samples to test that the median of the distribution of the difference between the two measurements on each individual is 0. For this we just apply **proc univariate** to the differences.

7.4   PROC TTEST

If we have independent samples $x_{11}, \dots, x_{1n_1}$ from the $N(\mu_1, \sigma_1)$ distribution and $x_{12}, \dots, x_{1n_2}$ from the $N(\mu_2, \sigma_2)$ distribution, where σ_1 and σ_2 are known then we can base inferences about the difference of the means $\mu_1 - \mu_2$ on the z statistic given by

$$z = \frac{\bar{x}_1 - \bar{x}_2 - (\mu_1 - \mu_2)}{\sqrt{\frac{\sigma_1^2}{n_1} + \frac{\sigma_2^2}{n_2}}}.$$

Under these assumptions z has a $N(0, 1)$ distribution. Therefore a $1 - \alpha$ confidence interval for $\mu_1 - \mu_2$ is given by

$$\bar{x}_1 - \bar{x}_2 \pm \sqrt{\frac{\sigma_1^2}{n_1} + \frac{\sigma_2^2}{n_2}} z^*$$

where z^* is the $1 - \alpha/2$ percentile of the $N(0, 1)$ distribution. Further, we can test $H_0 : \mu = \mu_0$ against the alternative $H_a : \mu \neq \mu_0$ by computing the P-value $P(|Z| > |z_0|) = 2P(Z > z_0)$, where Z is distributed $N(0, 1)$ and z_0 is the observed value of the z statistic. These inferences are also appropriate without normality provided n_1 and n_2 are large and we have reasonable values for σ_1 and σ_2. These inferences are easily carried out using SAS statements we have already discussed.

   In general, however, we will not have available suitable values of σ_1 and σ_2 or large samples and will have to use the two-sample analogs of the single-sample t procedures just discussed. This is acceptable provided, of course, that we have checked that both samples are from normal distributions and have agreed that it is reasonable to assume they are. These procedures are based on the two-sample t statistic given by

$$t = \frac{\bar{x}_1 - \bar{x}_2 - (\mu_1 - \mu_2)}{\sqrt{\frac{s_1^2}{n_1} + \frac{s_2^2}{n_2}}}$$

where we have replaced the population standard deviations by their sample estimates, when we don't assume equal population variances, and on

$$t = \frac{(\bar{x}_1 - \bar{x}_2)}{\left[s^2 \left(\frac{1}{n_1} + \frac{1}{n_2} \right) \right]^{\frac{1}{2}}}$$

where

$$s^2 = \frac{(n_1 - 1) s_1^2 + (n_2 - 1) s_2^2}{n_1 + n_2 - 2}$$

when we do assume equal population variances. Under the assumption of equal variances, $t \sim Student(n_1 + n_2 - 2)$. When we can't assume equal variances, the exact distribution of t does not have a convenient form, but of course we can always simulate its distribution. Actually it is typical to use an approximation to the distribution of this statistic based on a Student distribution. See the discussion on this topic in IPS.

The **proc ttest** procedure carries out the two-sample t test that two means are equal. The procedure *assumes* that the samples are from normal distributions. The following statements are available in this procedure.

proc ttest *options*;
var *variables*;
class *variable*;
by *variables*;

The option

data=*SASdataset*

can be used with the **proc ttest** statement, where *SASdataset* is a SAS data set containing the variables.

TTEST PROCEDURE

Variable: X

GENDER	N	Mean	Std Dev	Std Error	Variances	T	DF	Prob>|T|
f	4	2.77500000	1.13247517	0.56623758	Unequal	1.8046	4.5	0.1377
m	3	1.60000000	0.55677644	0.32145503	Equal	1.6275	5.0	0.1646

For H0: Variances are equal, F' = 4.14   DF = (3,2)   Prob>F' = 0.4015

Figure 7.1: Output from **proc ttest**.

The program

```
data one;
input sex $ x;
cards;
m 1.1
m 2.2
m 1.5
f 2.6
f 1.8
f 4.4
f 2.3
proc ttest data=one;
class sex;
```

splits the SAS data set **one** into two groups by the values of the variable **sex** and produces the output shown in Figure 7.1, which gives the values of both two-sample t statistcs and the P-values for testing $H_0 : \mu_1 = \mu_2$ against the alternative $H_a : \mu_1 \neq \mu_2$. In this case neither test rejects H_0.

A **class** statement must appear, and the class variable can be numeric or character but must assume exactly two values. The procedure also outputs the F test for testing the null hypothesis that the two population variances are equal; see Section II.7.5. The **by** and **var** statements work as with other procedures.

Simulation can be used to approximate the power of the two-sample t test. Note that in this case we must specify the difference $\mu_1 - \mu_2$ as well as σ_1 and σ_2. See Exercise 8 for further details.

7.5 F **Distribution**

If X_1 is distributed $Chisquare(k_1)$ independent of X_2 distributed $Chisquare(k_2)$, then

$$F = \frac{X_1/k_1}{X_2/k_2}$$

is distributed according to the $F(k_1, k_2)$ distribution. The value k_1 is called the *numerator degrees of freedom* and the value k_2 is called the *denominator degrees of freedom*. SAS functions assist in carrying out computations for this distribution.

The values of the density curve for the $F(k_1, k_2)$ distribution can be plotted as in Section II.6.4. The **probf** and **finv** functions are available to obtain values of the $F(k_1, k_2)$ cumulative distribution function and inverse distribution function, respectively. For example,

```
data;
p=probf(12,4,5);
x=finv(.65,3,12);
put 'p= ' p 'x= ' x;
run;
```

calculates the value of the $F(4, 5)$ distribution function at 12 in p, calculates the .65 quantile of the $F(3, 12)$ distribution in x, and writes

```
p= 0.9910771094 x= 1.2045020058
```

in the Log window. To generate from the $F(k_1, k_2)$ distribution, we generate $X_1 \sim Chisquare(k_1)$, $X_2 \sim Chisquare(k_2)$ and then form F as before.

A number of applications of the F distribution arise later in the book but we mention one here. In particular, if $x_{11}, \ldots, x_{1n_1}$ is a sample from the $N(\mu_1, \sigma_1)$ distribution and $x_{12}, \ldots, x_{1n_2}$ a sample from the $N(\mu_2, \sigma_2)$ distribution, then

$$F = \frac{s_1^2/\sigma_1^2}{s_2^2/\sigma_2^2}$$

is known to follow a $F(n_1 - 1, n_2 - 1)$. As explained in IPS, this fact is used as a basis for inference about the ratio σ_1/σ_2, i.e., confidence intervals and tests of significance and in particular testing for equality of variances between the samples. Because of the nonrobustness of these inferences to small deviations from normality, the inferences are not recommended.

7.6   Exercises

When the data for an exercise come from an exercise in IPS, the IPS exercise number is given in parentheses ( ). All computations in these exercises are to be carried out using SAS, and the exercises are designed to reinforce your understanding of the SAS material in this chapter. More generally, you should use SAS to do all the computations and plotting required for the problems in IPS.

1. The formula for the density curve of the *Student*(k) is given by

$$f(x) = \frac{\Gamma\left(\frac{\lambda+1}{2}\right)}{\Gamma\left(\frac{\lambda}{2}\right)\Gamma\left(\frac{1}{2}\right)} \left(1 + \frac{x^2}{\lambda}\right)^{-\frac{\lambda+1}{2}} \left(\frac{1}{\sqrt{\lambda}}\right)$$

   for $-\infty < x < \infty$. Using the method of Section II.6.4, plot the *Student*(k) density curve for $k = 1, 2, 10, 30$ and the $N(0,1)$ density curve at 1000 equispaced points in the interval $(-10, 10)$. Compare the plots.

2. Make a table of the values of the cumulative distribution function of the *Student*(k) distribution for $k = 1, 2, 10, 30$ and the $N(0,1)$ distribution at points $-10, -5, -3, -1, 0, 1, 3, 5, 10$. Comment on the values.

3. Make a table of the values of the inverse cumulative distribution function of the *Student*(k) distribution for $k = 1, 2, 10, 30$ and the $N(0,1)$ distribution at the points .0001, .001, .01, .1, .25, .5. Comment on the values.

4. Simulate $N = 1000$ values from Z distributed $N(0,1)$ and X distributed *Chisquare*(3) and plot a histogram of $T = Z/\sqrt{X/3}$ using the midpoints $-10, -9, \ldots, 9, 10$. Generate a sample of $N = 1000$ values directly from the *Student*(3) distribution, plot a histogram with the same midpoints, and compare the two histograms.

5. Carry out a simulation with $N = 1000$ to verify that the 95% confidence interval based on the t statistic covers the true value of the mean 95% of the time when taking samples of size 5 from the $N(4,2)$ distribution.

6. Generate a sample of 50 from the $N(10,2)$ distribution. Compare the 95% confidence intervals obtained via the t statistic and the z statistic using the sample standard deviation as an estimate of σ.

7. Carry out a simulation with $N = 1000$ to approximate the power of the t-test at $\mu_1 = 1, \sigma_1 = 2$ for testing $H_0 : \mu = 0$ versus the alternative $H_a : \mu \neq 0$ at level $\alpha = .05$ based on a sample of five from the normal distribution.

8. Carry out a simulation with $N = 1000$ to approximate the power of the two-sample t test at $\mu_1 = 1, \sigma_1 = 2, \mu_2 = 2, \sigma_1 = 3$ for testing $H_0 : \mu_1 - \mu_2 = 0$ versus the alternative $H_a : \mu_1 - \mu_2 \neq 0$ at level $\alpha = .05$ based on a sample of five from the $N(\mu_1, \sigma_1)$ distribution and a sample of eight from the $N(\mu_2, \sigma_2)$ distribution. Use the conservative rule when choosing the degrees of freedom for the approximate test: choose the smaller of $n_1 - 1$ and $n_2 - 1$.

9. If Z is distributed $N(\mu, 1)$ and X is distributed $Chisquare(k)$ independent of Z, then

$$Y = \frac{Z}{\sqrt{X/k}}$$

is distributed according to a *noncentral Student(k)* distribution with noncentrality μ. Simulate samples of $N = 1000$ from this distribution with $k = 5$ and $\mu = 0, 1, 5, 10$. Plot the samples in histograms with midpoints $-20, -19, \ldots, 19, 20$ and compare the plots.

10. The density curve of the $F(k_1, k_2)$ distribution is given by

$$f(x) = \frac{\Gamma\left(\frac{k_1+k_2}{2}\right)}{\Gamma\left(\frac{k_1}{2}\right)\Gamma\left(\frac{k_2}{2}\right)} \left(\frac{k_1}{k_2}x\right)^{\frac{k_1}{2}-1} \left(1 + \frac{k_1}{k_2}x\right)^{-\frac{k_1+k_2}{2}} \left(\frac{k_1}{k_2}\right)$$

for $x > 0$. For $k_1 = 1, 5, 10$ and $k_2 = 1, 5, 10$, plot the densities on $(0, 30)$.

Chapter 8

Inference for Proportions

This chapter is concerned with inference methods for a proportion p and for the comparison of two proportions p_1 and p_2. Proportions arise from measuring a binary-valued categorical variable on population elements such as gender in human populations. For example, p might be the proportion of females in a given population or we might want to compare the proportion p_1 of females in population 1 with the proportion p_2 of females in population 2. The need for inference arises as we base our conclusions about the values of these proportions on samples from the populations rather than every element in populations. For convenience, we denote the values assumed by the binary categorical variables as 1 and 0, where 1 indicates the presence of a characteristic and 0 indicates its absence.

8.1 Inference for a Single Proportion

Suppose $x_1, \ldots, x_n$ is a sample from a population where the variable is the presence or absence of some trait, indicated by a 1 or 0, respectively. Let $\hat{p}$ be the proportion of 1's in the sample. This is the estimate of the true proportion p. For example, the sample could arise from coin tossing where 1 denotes heads and 0 tails and $\hat{p}$ is the proportion of heads while p is the probability of heads. If the population we are sampling from is finite, then strictly speaking the sample elements are not independent. But if the population size is large relative to the sample size n, then independence is a reasonable approximation; independence is necessary for the methods of this chapter. So we will consider $x_1, \ldots, x_n$ as a sample from the *Bernoulli*(p) distribution.

The standard error of the estimate $\hat{p}$ is $\sqrt{\hat{p}(1 - \hat{p})/n}$, and since $\hat{p}$ is an average, the central limit theorem gives that

$$z = \frac{\hat{p} - p}{\sqrt{\frac{\hat{p}(1-\hat{p})}{n}}}$$

is approximately $N(0, 1)$ for large n. This leads to the approximate $1 - \alpha$ confidence interval given by $\hat{p} \pm \sqrt{\hat{p}(1 - \hat{p})/n}z^*$, where z^* is the $1 - \alpha/2$ percentile of the $N(01)$ distribution. This interval can be easily computed using SAS commands. For example, in Example 8.2 in IPS the probability of heads was estimated by Count Buffon as $\hat{p} = .5069$ on the basis of a sample of $n = 4040$ tosses. The statements

```
data;
p=.5069;
std=sqrt(p*(1-p)/4040);
z=probit(.95);
cl=p-std*z;
cu=p+std*z;
put '90% confidence interval for p is (' cl ',' cu ')';
run;
```

compute the approximate 90% confidence interval

```
90% confidence interval for p is (0.4939620574,0.5198379426)
```

which is printed in the Log window.

To test a null hypothesis $H_0 : p = p_0$, we make use of the fact that under the null hypothesis the statistic

$$z = \frac{\hat{p} - p_0}{\sqrt{\frac{p_0(1-p_0)}{n}}}$$

is approximately $N(0, 1)$. To test $H_0 : p = p_0$ versus $H_a : p \neq p_0$, we compute $P(|Z| > |z|) = 2P(Z > |z|)$, where Z is distributed $N(0, 1)$. For example, in Example 8.2 of IPS suppose we want to test $H_0 : p = .5$ versus $H_a : p \neq .5$. Then the statements

```
data;
p=.5069;
p0=.5;
std=sqrt(p0*(1-p0)/4040);
z=abs((p-p0)/std);
pval=2*(1-probnorm(z));
put 'z-statistic =' z 'P-value = ' pval;
run;
```

compute the value of the z statistic and the P-value of this two-sided test to be

```
z-statistic =0.8771417217 P-value = 0.3804096656
```

which is printed in the Log window. The formulas provided in IPS for computing the P-values associated with one-sided tests are also easily implemented in SAS.

8.2 Inference for Two Proportions

Suppose that $x_{11}, \ldots, x_{n_1 1}$ is a sample from population 1 and $x_{12}, \ldots, x_{n_2 2}$ is a sample from population 2 where the variable is measuring the presence or absence of some trait by a 1 or 0 respectively. We assume then that we have a sample of n_1 from the $Bernoulli(p_1)$ distribution and a sample of n_2 from the $Bernoulli(p_2)$ distribution. Suppose we want to make inferences about the difference in the proportions $p_1 - p_2$. Let $\hat{p}_i$ be the proportion of 1's in the $i - th$ sample.

The central limit theorem gives that

$$z = \frac{\hat{p}_1 - \hat{p}_2 - (p_1 - p_2)}{\sqrt{\frac{\hat{p}_1(1-\hat{p}_1)}{n_1} + \frac{\hat{p}_2(1-\hat{p}_2)}{n_2}}}$$

is approximately $N(0,1)$ for large n_1 and n_2. This leads to the approximate $1 - \alpha$ confidence interval given by

$$\hat{p}_1 - \hat{p}_2 \pm \sqrt{\frac{\hat{p}_1(1-\hat{p}_1)}{n_1} + \frac{\hat{p}_2(1-\hat{p}_2)}{n_2}} z^*$$

where z^* is the $1 - \alpha/2$ percentile of the $N(01)$ distribution. We can compute this interval using SAS commands just as we did for a confidence interval for a single proportion in Section II.8.1.

To test a null hypothesis $H_0 : p_1 = p_2$, we use the fact that under the null hypothesis the statistic

$$z = \frac{\hat{p}_1 - \hat{p}_2}{\sqrt{\hat{p}(1 - \hat{p})\left(\frac{1}{n_1} + \frac{1}{n_2}\right)}}$$

is approximately $N(0, 1)$ for large n_1 and n_2, where

$$\hat{p} = (n_1\hat{p}_1 + n_2\hat{p}_2) / (n_1 + n_2)$$

is the estimate of the common value of the proportion when the null hypothesis is true. To test $H_0 : p_1 = p_2$ versus $H_a : p_1 \neq p_2$, we compute $P(|Z| > |z|) = 2P(Z > |z|)$ where Z is distributed $N(0, 1)$. For example, in Example 8.9 of IPS, suppose we want to test $H_0 : p_1 = p_2$ versus $H_a : p_1 \neq p_2$, where $n_1 = 7180, \hat{p}_1 = .227, n_2 = 9916, \hat{p}_2 = .170$. Then the statements

```
data;
p1=.227;
p2=.170;
n1=7180;
n2=9916;
p=(n1*p1+n2*p2)/(n1+n2);
std=sqrt(p*(1-p)*(1/n1+1/n2));
z=abs((p1-p2)/std);
pval=2*(1-probnorm(z));
put 'z-statistic =' z 'P-value = ' pval;
run;
```

compute the z statistic and the P-value and print

```
z-statistic =9.3033981753 P-value = 0
```

in the Log window. Here the P-value equals 0, so we would definitely reject.

Approximate power calculations can be carried out by simulating N pairs of values from the $Binomial(n_1, p_1)$ and $Binomial(n_2, p_2)$ distributions. For example, the statements

```
data;
seed=43567;
p1=.3;
p2=.5;
n1=40;
n2=50;
power=0;
do i=1 to 1000;
k1=ranbin(seed,n1,p1);
p1hat=k1/n1;
k2=ranbin(seed,n2,p2);
p2hat=k2/n2;
phat=(n1*p1hat+n2*p2hat)/(n1+n2);
std=sqrt(phat*(1-phat)*(1/n1+1/n2));
z=abs((p1hat-p2hat)/std);
pval=2*(1-probnorm(z));
if pval le .05 then
power=power+1;
end;
power=power/1000;
stderr=sqrt(power*(1-power)/1000);
put 'Approximate power (standard error) at
      p1=.3, p2=.5 n1=40 and n2=50 equals';
put power '(' stderr ')';
run;
```

simulate generating 1000 samples of sizes 40 and 50 from the *Bernoulli*(.3) and *Bernoulli*(.5), respectively, and then testing whether or not the proportions are equal. The null hypothesis of equality is rejected whenever the *P*-value is less than or equal to .05. The variable **power** records the proportion of rejections in the simulations. The approximate power and its standard error are given by

```
0.514 (0.015805189)
```

which is printed in the Log window.

8.3   Exercises

When the data for an exercise come from an exercise in IPS, the IPS exercise number is given in parentheses ( ). All computations in these exercises are to be carried out using SAS, and the exercises are designed to reinforce your understanding of the SAS material in this chapter. More generally, you should use SAS to do all the computations and plotting required for the problems in IPS.

   Don't forget to quote standard errors for any approximate probabilities you quote in the following problems.

1. Carry out a simulation with the $Binomial(40, .3)$ distribution to assess the coverage of the 95% confidence interval for a single proportion.

2. The accuracy of a confidence interval procedure can be assessed by computing *probabilities of covering false values*. Approximate the probabilities of covering the values $.1, .2, \ldots, .9$ for the 95% confidence interval for a single proportion when sampling from the $Binomial(20, .5)$ distribution.

3. Approximate the power of the two-sided test for testing $H_0 : p = .5$ at level $\alpha = .05$ at the points $n = 100, p = .1, \ldots, 9$ and plot the power curve.

4. Carry out a simulation with the $Binomial(40, .3)$ and the $Binomial(50, .4)$ distribution to assess the coverage of the 95% confidence interval for a difference of proportions.

5. Approximate the power of the two-sided test for testing $H_0 : p_1 = p_2$ versus $H_a : p_1 \neq p_2$ at level $\alpha = .05$ at $n_1 = 40, p_1 = .3, n_2 = 50, p_2 = .1, \ldots, 9$ and plot the power curve.

Chapter 9

Inference for Two-way Tables

In this chapter inference methods are discussed for comparing the distributions of a categorical variable for a number of populations and for looking for relationships amongst a number of categorical variables defined on a single population. The *chi-square test* is the basic inferential tool, and it can be carried out in SAS via the **proc freq** statement.

9.1   PROC FREQ with Nontabulated Data

You should recall or reread the discussion of the **proc freq** statement in Section II.1.1, as we mention here only the additional features related to carrying out the chi-square test. For example, suppose that for 100 observations in a SAS data set **one** we have a categorical variable **x1** taking the values 0 and 1 and a categorical variable **x2** taking the values 0, 1, and 2. Then the statements

```
proc freq data=one;
tables x1*x2;
run;
```

record the counts in the six cells of a table with **x1** indicating row and **x2** indicating column (Figure 9.1). The variable **x1** could be indicating a population

```
                        TABLE OF X1 BY X2

     X1          X2

     Frequency|
     Percent  |
     Row Pct  |
     Col Pct  |       0|       1|       2|  Total

          0   |       7|      12|       3|     22
              |    7.00|   12.00|    3.00|  22.00
              |   31.82|   54.55|   13.64|
              |   19.44|   24.00|   21.43|
     ---------+--------+--------+--------+
          1   |      29|      38|      11|     78
              |   29.00|   38.00|   11.00|  78.00
              |   37.18|   48.72|   14.10|
              |   80.56|   76.00|   78.57|
     ---------+--------+--------+--------+
     Total           36       50       14      100
                  36.00    50.00    14.00   100.00
```

Figure 9.1: Two-way table produced by **proc freq**.

with **x2** a categorical variable defined on each population (or conversely), or both variables could be defined on a single population.

There is no relationship between two random variables — the variables are *independent* — if and only if the conditional distributions of x2 given **x1** are all the same. In terms of the table this means comparing the two distributions (.3182, .5455, .1364) and (.3718, .4872, .1410). Alternatively we can compare the conditional distributions of x1 given x2, i.e., compare the three distributions (.1944, .8056), (.2400, .7600) and (.2143, .7857). Of course, there will be differences in these conditional distributions simply due to sampling error. Whether or not the differences are significant is assessed by conducting a chi-square test, which can be carried out using the **chisq** option to the **tables** statement. The SAS statements

```
proc freq data=one;
tables x1*x2 /chisq cellchi2;
run;
```

produce the results shown in Figure 9.2. The table is the same as that in Figure 9.1 with the exception that the **expected** option causes the expected value of each cell to be printed and the **cellchi2** option causes the contribution of each cell to the chi-square statistic

$$\chi^2 = \sum_{\text{cell}} \frac{(\text{observed count in cell} - \text{expected count in cell})^2}{\text{expected count in cell}}$$

to be printed in the corresponding cell; i.e., in the (i, j)-th cell the value

$$\frac{(\text{observed count in cell} - \text{expected count in cell})^2}{\text{expected count in cell}}$$

is printed as `Cell Chi-square`. In this case the chi-square statistic takes the value .256 and the P-value is .880, which indicates that there is no evidence of a difference among the conditional distributions, that is, no evidence against the statistical independence of x1 and x2. The P-value of the chi-square test is obtained by computing the probability

$$P(Y > \chi^2)$$

where Y follows a *Chisquare* (k) distribution based on an appropriate degrees of freedom k as determined by the table and the model being fitted. When the table has r rows and c columns and we are testing for independence, then $k = (r - 1)(c - 1)$. This is an approximate distribution result. Recall that the *Chisquare* (k) distribution was discussed in Section II.6.4.

It is possible to cross-tabulate more than two variables and to test for pairwise statistical independence among the variables using the **chisq** option. For example, if there are 3 categorical variables x1, x2, and x3 in SAS data set **one**, then

```
proc freq data=one;
table x1*x2*x3/chisq ;
run;
```

causes a two-way table of x2 by x3 to be created for each value of x1 and a chi-square test to be carried out for each two-way table.

```
X1                X2

Frequency
Expected
Cell Chi-Square
Percent
Row Pct
Col Pct                        0|        1|        2|   Total

        0          7        12         3       22
                 7.92        11      3.08
               0.1069    0.0909    0.0021
                 7.00     12.00      3.00    22.00
                31.82     54.55     13.64
                19.44     24.00     21.43

        1         29        38        11       78
                28.08        39     10.92
               0.0301    0.0256    0.0006
                29.00     38.00     11.00    78.00
                37.18     48.72     14.10
                80.56     76.00     78.57

    Total         36        50        14      100
               36.00     50.00     14.00   100.00

        STATISTICS FOR TABLE OF X1 BY X2

    Statistic                  DF    Value      Prob

    Chi-Square                  2    0.256     0.880
    Likelihood Ratio Chi-Square 2    0.258     0.879
    Mantel-Haenszel Chi-Square  1    0.090     0.764
    Phi Coefficient                  0.051
    Contingency Coefficient          0.051
    Cramer's V                       0.051

    Sample Size = 100
```

Figure 9.2: Two-way table produced by **proc freq** with the **chisq**, **expected** and **cellchi2** option to the **tables** statement.

9.2 PROC FREQ with Tabulated Data

If the data come to you already tabulated, then you must use the **weight** statement in **proc freq** together with the **chisq** option in the **tables** statement to compute the chi-square statistic. For example, consider the data in the table of Example 9.2 of IPS. Then the program

```
data one;
input binge $ gender $ count;
cards;
yes men 1630
yes women 1684
no men 5550
no women 8232
proc freq data=one;
weight count;
tables binge*gender/chisq;
run;
```

uses the **weight** statement to record the counts in each of the cells of the

2×2 table formed by **gender** and **binge** in the variable **counts**. The output for this program is shown in Figure 9.3. We see that the chi-square test for this table gives a *P*-value of .001, so we would reject the null hypothesis of no relationship between the variables **gender** and **binge**.

```
GENDER     BINGE

Frequency|
Percent  |
Row Pct  |
Col Pct  |men     |women   | Total

no           5550     8232    13782
            32.46    48.15    80.62
            40.27    59.73
            77.30    83.02

yes          1630     1684     3314
             9.53     9.85    19.38
            49.19    50.81
            22.70    16.98

Total        7180     9916    17096
            42.00    58.00   100.00


        STATISTICS FOR TABLE OF GENDER BY BINGE

Statistic                        DF      Value        Prob

Chi-Square                        1     87.172        0.001
Likelihood Ratio Chi-Square       1     86.396        0.001
Continuity Adj. Chi-Square        1     86.806        0.001
Mantel-Haenszel Chi-Square        1     87.167        0.001
Fisher's Exact Test (Left)                          8.80E-21
                    (Right)                            1.000
                    (2-Tail)                        1.63E-20
Phi Coefficient                        -0.071
Contingency Coefficient                 0.071
Cramer's V                             -0.071

Sample Size = 17096
```

Figure 9.3: The chi-square test on the data in Example 9.2 of IPS. This illustrates the use of **proc freq** with already tabulated data.

9.3 Exercises

When the data for an exercise come from an exercise in IPS, the IPS exercise number is given in parentheses ( ). All computations in these exercises are to be carried out using SAS, and the exercises are designed to reinforce your understanding of the SAS material in this chapter. More generally, you should use SAS to do all the computations and plotting required for the problems in IPS.

1. Use SAS to directly compute the expected frequencies, standardized residuals, chi-square statistic, and *P*-value for the hypothesis of independence in the table of Example 9.8 in IPS.

2. (9.17) Plot bar charts of the conditional distributions. Make sure you use the same scale on each plot so that they are comparable.

3. Suppose we have a discrete distribution on the integers $1, \ldots, k$ with probabilities $p_1, \ldots, p_k$. Further suppose we take a sample of n from this distribution and record the counts $f_1, \ldots, f_k$ where f_i records the number of times we observed i. Then it can be shown that

$$P(f_1 = n_1, \ldots, f_k = n_k) = \left(\frac{n!}{n_1! \cdots n_k!} \right) \left(p_1^{n_1} \cdots p_k^{n_k} \right)$$

when the n_i are nonnegative integers that sum to n. This is called the *Multinomial*$(n, p_1, \ldots, p_k)$ distribution, and it is a generalization of the *Binomial*(n, p) distribution. It is the relevant distribution for describing the counts in cross-tabulations. For $k = 4, p_1 = p_2 = p_3 = p_4 = .25, n = 3$ calculate these probabilities and verify that it is a probability distribution. Recall that the **gamma** function (see Appendix A) can be used to evaluate factorials such as $n!$ and also $0! = 1$.

4. Calculate $P(f_1 = 3, f_2 = 5, f_3 = 2)$ for the *Multinomial*$(10, .2, .5, .3)$ distribution.

5. Generate (f_1, f_2, f_3) from the *Multinomial*$(1000, .2, .4, .4)$ distribution. Hint: Generate a sample of 1000 from the discrete distribution on 1, 2, 3 with probabilities .2, .4 , .4 respectively.

Chapter 10

Inference for Regression

This chapter deals with inference for the simple linear model. The procedure **proc reg** for the fitting of this model was discussed in Section II.2.1.3, and this material should be recalled or reread at this point. Here we discuss a number of additional features available with **proc reg** and present an example.

10.1   PROC REG

The **proc reg** procedure fits the model $y = \beta_0 + \beta_1 x + \epsilon$, where $\beta_0, \beta_1 \in R^1$ are unknown and to be estimated, and $\epsilon \sim N(0, \sigma)$ with $\sigma \in [0, \infty)$ unknown and to be estimated. We denote the least-squares estimates of β_0 and β_1 by b_0 and b_1, respectively, where they are based on the observed data $(x_1, y_1), \ldots, (x_n, y_n)$. We also estimate the standard deviation σ by s which equals the square root of the MSE (mean-squared error) for the regression model.

Following are some of the statements that can appear with this procedure. We discuss here only features that were not mentioned in Section II.2.1.3.

proc reg *options*;
model *dependent=independent /options*;
by *variables*;

freq *variable*;
id variable;
var *variables*;
weight *variable*;
plot *yvariable*xvariable = 'symbol' /options*;
output out = *SAS-dataset* **keyword** = *names*;
test *linear combination = value /option*;

The **test** statement is used to test null hypotheses of the form $H_0 : l_0\beta_0 + l_1\beta_1 = c$ versus $H_a : l_0\beta_0 + l_1\beta_1 \neq c$. The statement takes the form

test $l_0 * intercept + l_1 * variable = c/$**print**;

where *variable* is the predictor variable. The **print** option causes some intermediate calculations to be output and can be deleted if the calculations are not needed. Actually we can use this statement to construct predictions for the response at given settings of the predictor variables. This is illustrated in Section 10.2.

Following are some of the options that may appear in the **proc reg** statement.

data = *SASdataset*
corr
simple
noprint

Following are some of the options that may appear in the **model** statement.

cli prints 95% confidence limits for predicted values for observations.

clm prints 95% confidence limits for expected values for observations.

collin requests a collinearity analysis (see reference 4 in Appendix E for discussion).

covb prints the covariance matrix for the least-squares estimates.

influence requests an influence analysis of each observation, the (ordinary) residual, the leverage, rstudent (deleted studentized residual), covratio, dffits and dfbetas are all printed (see reference 4 in Appendix E for definitions).

noint causes the model to be fit without an intercept term β_0.

p prints the predicted values for the observations and the (ordinary = observation − prediction) residuals.

r requests a residual analysis, the predicted values, the standard errors of the predicted values, the (ordinary = observation − prediction) residuals and their standard errors, the studentized residuals (ordinary residuals divided by their standard errors here) and Cook's D are all printed (see reference 4 in Appendix E for more details). Note that the *studentized residuals* are often referred to as the *standardized residuals*.

The following **keywords** may appear in the **output** statement. The values of *names* are any valid SAS names, one for each model fit.

l95 = *names* lower 95% bound for prediction of observation.

u95 = *names* upper 95% bound for prediction of observation.

l95m = *names* lower 95% bound for expectation of observation.

u95m = *names* upper 95% bound for expectation of observation.3

p = *names* predicted values of observations.

r = *names* residuals (ordinary) of observations.

stdi = *names* standard error of predicted value.

stdr = *names* standard error of residual (ordinary).

student = *names* studentized (standardized residuals.

The **overlay** option may appear in the **plot** statement. The **overlay** option causes multiple scatterplots to be plotted on the same set of axes.

10.2 Example

We illustrate the use of **proc reg** with Example 10.8 in IPS. We have four data points

$$
\begin{aligned}
(x_1, y_1) &= (1966, 73.1) \\
(x_2, y_2) &= (1976, 88.0) \\
(x_3, y_3) &= (1986, 119.4) \\
(x_4, y_4) &= (1996, 127.1)
\end{aligned}
$$

where x is year and y is yield in bushels per acre. Suppose we give x the name `year` and y the name `yield` and place this data in the SAS data set `ex108`. Then the program

```
data ex108;
```

```
input year yield;
cards;
1966 73.1
1976 88.0
1986 119.4
1996 127.1
proc reg data=ex108;
model yield=year/p;
output out=regstuff student=stdresid;
proc univariate data=regstuff plot;
var stdresid;
run;
```

produces the output from **proc reg** shown in Figure 10.1. This gives the least-squares line as $y = -3729.354 + 1.934x$. The standard error of $b_0 = -3729.4$ is 606.6, the standard error of $b_1 = 1.934$ is 0.3062, the t statistic for testing $H_0 : \beta_0 = 0$ versus $H_a : \beta_0 \neq 0$ is -6.148 with P-value 0.0255, and the t statistic for testing $H_0 : \beta_1 = 0$ versus $H_a : \beta_1 \neq 0$ is 6.316 with P-value 0.0242. The estimate of σ is $s = 6.847$, and the squared correlation is $R^2 = .9523$, indicating that 95% of the observed variation in y is explained by the changes in x. The analysis of variance table indicates that the F statistic for testing $H_0 : \beta_1 = 0$ versus $H_a : \beta_1 \neq 0$ is 39.892 with P-value 0.0242 and the MSE (mean-squared error) is 46.881. The predicted value at $x = 1996$ is 130.9, and so on.

In Figure 10.2 some partial output from **proc univariate** is shown. In particular this gives a stem-and-leaf, boxplot, and normal probability plot of the standardized residuals. These plots don't show any particular grounds for concern, but of course there is very little data.

Now suppose we want to predict the value of the response at $x = 2000$. Then the program

```
data ex108;
input year yield;
cards;
1966 73.1
1976 88.0
1986 119.4
1996 127.1
proc reg data=ex108;
```

```
Model: MODEL1
Dependent Variable: YIELD

                              Analysis of Variance

                                    Sum of          Mean
             Source        DF      Squares        Square     F Value    Prob>F

             Model          1    1870.17800    1870.17800     39.892    0.0242
             Error          2      93.76200      46.88100
             C Total        3    1963.94000

                 Root MSE        6.84697      R-square      0.9523
                 Dep Mean      101.90000      Adj R-sq      0.9284
                 C.V.           6.71930

                              Parameter Estimates

                       Parameter      Standard    T for H0:
             Variable  DF  Estimate       Error  Parameter=0   Prob > |T|

             INTERCEP   1 -3729.354000  606.60337646      -6.148       0.0255
             YEAR       1     1.934000    0.30620581       6.316       0.0242

                             Dep Var     Predict
                       Obs     YIELD       Value    Residual

                        1     73.1000    72.8900      0.2100
                        2     88.0000    92.2300     -4.2300
                        3      119.4      111.6       7.8300
                        4      127.1      130.9      -3.8100

   Sum of Residuals                    0
   Sum of Squared Residuals      93.7620
   Predicted Resid SS (Press)   323.4163
```

Figure 10.1: Output from **proc reg** for the example.

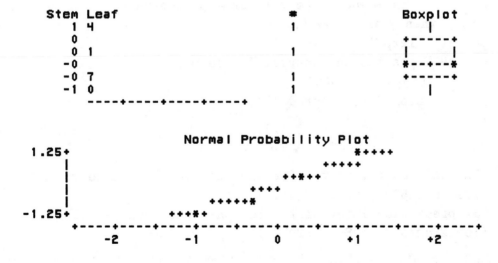

Figure 10.2: Part of the output from **proc univariate** for the example.

```
model yield=year;
test intercept + 2000*year=0/print;
run;
```

produces as part of its output that shown in Figure 10.3. The value Lb-c=138.646 is the prediction at year=2000. The standard error of this estimate is obtained by taking the square root of MSE*(L Ginv(X'X), L')which equals $\sqrt{46.881 * (0.972)} = 6.7504$ in this case. These ingredients can be used to construct confidence intervals for the expected value and predicted value. For example,

```
data;
est=138.646;
l=0.972;
s2=46.881;
stderr1=sqrt(s2*l);
stderr2=sqrt(1+s2*l);
t=tinv(.975,2);
clexp=est-stderr1*t;
cuexp=est+stderr1*t;
clpred=est-stderr2*t;
cupred=est+stderr2*t;
put '95% confidence interval for expected value
     when year = 2000';
put '(' clexp ',' cuexp ')';
put '95% prediction interval for response
     when year = 2000';
put '(' clpred ',' cupred ')';
run;
```

prints

```
95% confidence interval for expected value when year = 2000
(109.60123539 ,167.69076461 )
95% prediction interval for response when year = 2000
(109.28427028 ,168.00772972 )
```

in the Log window. Note the difference between the intervals. Also note that we have used the error degrees of freedom — namely, 2 — to determine the appropriate Student distribution to use in forming the intervals.

```
                    L Ginv(X'X) L'    Lb-c

                         0.972         138.646

            Inv(L Ginv(X'X) L')    Inv( )(Lb-c)

                    1.0288065844      142.6399177
```

Dependent Variable: YIELD
Test: Numerator: 19776.4540 DF: 1   F value: 421.8437
                     Denominator: 46.881 DF: 2   Prob>F: 0.0024

Figure 10.3: Output from the **test** statement in **proc reg** in the example.

10.3   Exercises

When the data for an exercise come from an exercise in IPS, the IPS exercise number is given in parentheses ( ). All computations in these exercises are to be carried out using SAS, and the exercises are designed to reinforce your understanding of the SAS material in this chapter. More generally, you should use SAS to do all the computations and plotting required for the problems in IPS.

1. For the x values $-3.0, -2.5, -2.0, \ldots, 2.5, 3.0$ and a sample of 13 from the error ϵ, where ϵ is distributed $N(0, 2)$, compute the values $y = \beta_0 + \beta_1 x + \epsilon = 1 + 3x + \epsilon$. Calculate the least-squares estimates of β_0 and β_1 and the estimate of σ^2. Repeat this example with five observations at each value of x. Compare the estimates from the two situations and their estimated standard deviations.

2. For the x values $-3.0, -2.5, -2.0, \ldots, 2.5, 3.0$ and a sample of 13 from the error ϵ, where ϵ is distributed $N(0, 2)$, compute the values $y = \beta_0 + \beta_1 x + \epsilon = 1 + 3x + \epsilon$. Plot the least-squares line. Now repeat your computations twice after changing the first y observation to 20 and then to 50 and make sure the scales on all the plots are the same. What effect do you notice?

3. For the x values $-3.0, -2.5, -2.0, \ldots, 2.5, 3.0$ and a sample of 13 from the error ϵ, where ϵ is distributed $N(0, 2)$, compute the values $y = \beta_0 + \beta_1 x + \epsilon = 1 + 3x + \epsilon$. Plot the standardized residuals in a normal

quantile plot against the fitted values and against the explanatory variable. Repeat your computations with the values of $y = 1 + 3x - 5x^2 + \epsilon$. Compare the residual plots.

4. For the x values $-3.0, -2.5, -2.0, \ldots, 2.5, 3.0$ and a sample of 13 from the error ϵ, where ϵ is distributed $N(0, 2)$, compute the values $y = \beta_0 + \beta_1 x + \epsilon = 1 + 3x + \epsilon$. Plot the standardized residuals in a normal quantile plot against the fitted values and against the explanatory variable. Repeat your computations but for ϵ use the values of a sample of 13 from the *Student*(1) distribution. Compare the residual plots.

5. For the x values $-3.0, -2.5, -2.0, \ldots, 2.5, 3.0$ and a sample of 13 from the error ϵ, where ϵ is distributed $N(0, 2)$, compute the values $y = \beta_0 + \beta_1 x + \epsilon = 1 + 3x + \epsilon$. Calculate the predicted values and the lengths of .95 confidence and prediction intervals for this quantity at $x = .1, 1.1, 2.1, 3.5, 5, 10$ and 20. Explain the effect you observe.

6. For the x values $-3.0, -2.5, -2.0, \ldots, 2.5, 3.0$ and a sample of 13 from the error ϵ, where ϵ is distributed $N(0, 2)$, compute the values $y = \beta_0 + \beta_1 x + \epsilon = 1 + 3x + \epsilon$. Calculate the least-squares estimates and their estimated standard deviations. Repeat your computations but for the x values use 12 values of -3 and one value of 3. Compare your results and explain them.

Chapter 11

Multiple Regression

SAS statement introduced in this chapter

proc glm

In this chapter we discuss *multiple regression* — we have a single numeric response variable y and $k > 1$ explanatory variables $x_1, \ldots, x_k$. The descriptions of the behavior of the **proc reg** procedure in Chapter 10 apply as well to this chapter. This chapter briefly introduces a much more comprehensive procedure for regression (**proc glm**). In Chapter 15 we show how a *logistic regression* — a regression where the response variable y is binary-valued — is carried out using SAS.

11.1 Example Using PROC REG

We consider a generated multiple regression example to illustrate the use of the **proc reg** command in this context. Suppose that $k = 2$ and

$$\begin{aligned} y &= \beta_0 + \beta_1 x + \beta_2 w + \epsilon \\ &= 1 + 2x + 3w + \epsilon \end{aligned}$$

where ϵ is distributed $N(0, \sigma)$ with $\sigma = 1.5$. We generated the data for this example. First we generated a sample of 16 from the $N(0, 1.5)$ distribution for the values of ϵ. These values, together with every possible combination of $x_1 = -1, -.5, .5, 1$ and $x_2 = -2, -1, 1, 2$, and with $\beta_0 = 1, \beta_1 = 2$ and

$\beta_2 = 3$, gave the values of y according to the above equation. We stored the values of (x, w, y) in the SAS data set `example`. We then proceed to analyze these data as if we didn't know the values of $\beta_0, \beta_1, \beta_2$, and σ. The program

```
proc reg data=example;
model y= x1 x2;
output out=resan student=stdres;
test intercept+.75*x1+0*x2=0/print;
proc univariate data=resan plot;
var stdres;
run;
```

produces the output given in Figures 11.1 and 11.2 from **proc reg** and partial output from **proc univariate** is given in Figure 11.3. The least-squares equation is given as $y = 1.861516 + 2..099219x_1 + 2.982794x_2$. For example, the estimate of β_1 is $b_1 = 2..099219$ with standard error 0.33257638, and the t statistic for testing $H_0 : \beta_1 = 0$ versus $H_a : \beta_1 \neq 0$ is 6.312 with P-value 0.0001. The estimate of σ is $s = 1.05170$ and $R^2 = .9653$. The analysis of variance table indicates that the F statistic for testing $H_0 : \beta_1 = \beta_2 = 0$ versus $H_a : \beta_1 \neq 0$ or $\beta_2 \neq 0$ takes the value 180.798 with P-value 0.0001, so we would definitely reject the null hypothesis. Also the MSE is given as 1.10607. The output from the **test** statement in Figure 11.2 gives the prediction at `x1=.75`, `x2=0` as 3.4359304135, and the standard error of this estimate is $\sqrt{(1.10607) * (0.11875)} = .3624166$, which can be used as in Section II.10.2 to form confidence intervals for the expected response and prediction intervals for the predicted response. Note that the error degrees of freedom here is 13, so we use the quantiles of the $Student(13)$ distribution in forming these intervals. Finally in Figure 11.3, we present a normal probability plot for these data based on the standardized residuals. As we might expect, the plot looks appropriate.

Model: MODEL1
Dependent Variable: Y

Analysis of Variance

Source	DF	Sum of Squares	Mean Square	F Value	Prob>F
Model	2	399.94959	199.97480	180.798	0.0001
Error	13	14.37892	1.10607		
C Total	15	414.32851			

Root MSE	1.05170	R-square	0.9653	
Dep Mean	1.86152	Adj R-sq	0.9600	
C.V.	56.49689			

Parameter Estimates

Variable	DF	Parameter Estimate	Standard Error	T for H0: Parameter=0	Prob > |T|
INTERCEP	1	1.861516	0.26292472	7.080	0.0001
X1	1	2.099219	0.33257638	6.312	0.0001
X2	1	2.982794	0.16628819	17.937	0.0001

Figure 11.1: Output from **proc reg** for the generated example.

L Ginv(X'X) L'	Lb-c
0.11875	3.4359304135

Inv(L Ginv(X'X) L')	Inv( )(Lb-c)
8.4210526316	28.934150851

Dependent Variable: Y

Test:					
Numerator:	99.4157	DF:	1	F value:	89.8819
Denominator:	1.106071	DF:	13	Prob>F:	0.0001

Figure 11.2: Output from the **test** statement in **proc reg** for the generated example.

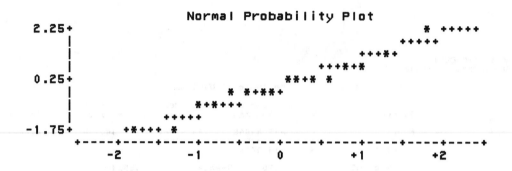

Figure 11.3: Normal probability plot obtained from **proc univariate** based on standardized residuals for generated data.

11.2   PROC GLM

SAS contains another regression procedure called **proc glm**. Whereas **proc reg** allows only quantitative variables for predictor variables, **proc glm** allows both quantitative and categorical predictor variables. Categorical variables are called *class variables* in SAS, and in **proc glm** they are identified in a **class** statement so that SAS knows to treat them appropriately. In Chapters 12 and 13 we discuss the situation where all the predictor variables are categorical, which can often be handled by another SAS procedure called **proc anova.** We elect, however, to use **proc glm** because it has certain advantages.

Suppose we analyze the data created in the SAS data set `example` in Section II.11.1. The program

```
proc glm data=example;
model y= x1 x2;
run;
```

produces the output shown in Figure 11.4. There are two tables after the analysis of variance table labeled Type I SS and Type III SS. In this case the entries are identical, so it doesn't matter which we use to test for the existence of individual terms in the model. These tables give the P-values for testing the hypotheses $H_0 : \beta_2 = 0$ versus $H_a : \beta_2 \neq 0$ and $H_0 : \beta_1 = 0$ versus $H_a : \beta_1 \neq 0$. We use the F statistics and the P-values given in the table to carry out these tests . We see that we reject $H_0 : \beta_2 = 0$ as the

```
|                              General Linear Models Procedure

Dependent Variable: Y

Source                    DF          Sum of Squares         Mean Square    F Value    Pr > F

Model                      2           399.94959046         199.97479523     180.80     0.0001

Error                     13            14.37891651           1.10607050

Corrected Total           15           414.32850696

                     R-Square              C.V.              Root MSE              Y Mean

                     0.965296            56.49689           1.0516988            1.86151628

Source                    DF              Type I SS          Mean Square    F Value    Pr > F
X1                         1            44.06719776          44.06719776      39.84     0.0001
X2                         1           355.88239269         355.88239269     321.75     0.0001

Source                    DF             Type III SS         Mean Square    F Value    Pr > F
X1                         1            44.06719776          44.06719776      39.84     0.0001
X2                         1           355.88239269         355.88239269     321.75     0.0001

                                                    T for H0:       Pr > |T|      Std Error of
Parameter                     Estimate           Parameter=0                       Estimate

INTERCEPT                   1.861516277              7.08           0.0001         0.26292472
X1                         2.099218849              6.31           0.0001         0.33257638
X2                         2.982793962             17.94           0.0001         0.16628819
```

Figure 11.4: Output from **proc glm** when applied to the generated example.

P-value is .0001. and similarly we reject $H_0 : \beta_1 = 0$.

In general Type I SS (sums of squares) and Type III SS will differ. Type I SS correspond to the drop in the Error SS entailed by adding the term corresponding to the row to the model, given that all the terms corresponding to rows above it are in the model. These SS are sometimes called *sequential sums of squares* and they are used when there is a natural sequence to testing whether or not terms are in the model. For example, when fitting a cubic $y = \beta_0 + \beta_1 x + \beta_2 x^2 + \beta_3 x^3 + \epsilon$, first test for the existence of the cubic term, then the quadratic term, then the linear term. Obviously the order in which we put variables into the model matters with these sequential tests except when we have balanced data. Type III sums of squares correspond to the drop in Error SS entailed by adding the term corresponding to the row to the model, given that all the terms corresponding to the remaining rows are in the model.

Following are some of the statements that can appear in **proc glm**.

proc glm;
class *variables*;
model *dependents = independents /options*;

by *variables*;
freq *variable*;
id *variable*;
weight *variable*;
output out =*SASdataset* **keyword** = *name*;

All of these work as they do in **proc reg** with similar options. The **class** statement simply lists all predictor variables that are categorical and appears before the **model** statement that references these predictors. We consider other features of **proc glm** in Chapters 12 and 13.

11.3 Exercises

When the data for an exercise come from an exercise in IPS, the IPS exercise number is given in parentheses ( ). All computations in these exercises are to be carried out using SAS, and the exercises are designed to reinforce your understanding of the SAS material in this chapter. More generally, you should use SAS to do all the computations and plotting required for the problems in IPS.

1. For the x_1 values $-3.0, -2.5, -2.0, \ldots, 2.5, 3.0$ and a sample of 13 from the error ϵ, where ϵ is distributed $N(0, 2)$, compute the values of $y = \beta_0 + \beta_1 x_1 + \beta_2 x_2 + \epsilon = 1 + 3x_1 + 5x_1^2 + \epsilon$. Calculate the least-squares estimates of β_0, β_1 and β_2 and the estimate of σ^2. Carry out the sequential F tests testing first for the quadratic term and then, if necessary, for the linear term.

2. For the x values $-3.0, -2.5, -2.0, \ldots, 2.5, 3.0$ and a sample of 13 from the error ϵ, where ϵ is distributed $N(0, 2)$, compute the values of $y = \beta_0 + \beta_1 x_1 + \beta_2 x_2 + \epsilon = 1 + 3\cos(x) + 5\sin(x) + \epsilon$. Calculate the least-squares estimates of β_0, β_1 and β_2 and the estimate of σ^2. Carry out the F test for any effect due to x. Are the sequential F tests meaningful here?

3. For the x_1 values $-3.0, -2.5, -2.0, \ldots, 2.5, 3.0$ and a sample of 13 from the error ϵ, where ϵ is distributed $N(0, 2)$, compute the values of $y = 1 + 3\cos(x) + 5\sin(x) + \epsilon$. Now fit the model $y = \beta_0 + \beta_1 x_1 + \beta_2 x_2 + \epsilon$ and plot the standardized residuals in a normal quantile plot and against each of the explanatory variables.

Chapter 12

One-way Analysis of Variance

This chapter deals with methods for making inferences about the relationship between a single numeric response variable and a single categorical explanatory variable. The basic inference methods are the one-way analysis of variance (ANOVA) and the comparison of means. For this we use the procedure **proc glm**. Other procedures for carrying out an ANOVA in SAS include **proc anova**. A disadvantage of **proc anova**, however, is that it must have *balanced data*; each cell formed by the cross classification of the explanatory variables has the same number of observations, and there are limitations with respect to residual analysis. Due to the importance of checking assumptions in a statistical analysis, we prefer to use **proc glm** and refer the reader to reference 3 in Appendix E for a discussion of **proc anova**.

We write the one-way ANOVA model as $x_{ij} = \mu_i + \epsilon_{ij}$, where $i = 1, \ldots, I$ indexes the levels of the categorical explanatory variable and $j = 1, \ldots, n_i$ indexes the individual observations at each level, μ_i is the mean response at the i-th level, and the errors ϵ_{ij} are a sample from the $N(0, \sigma)$ distribution. Based on the observed x_{ij}, we want to make inferences about the unknown values of the parameters $\mu_1, \ldots, \mu_I, \sigma$.

12.1   Example

We analyze the data of Example 12.6 in IPS using **proc glm**. For this
example there are $I = 3$ levels corresponding to the values `Basal`, `DRTA`, and
`Strat`, and $n_1 = n_2 = n_3 = 22$. Suppose we have the values of the x_{ij} in
`score` and the corresponding values of the categorical explanatory variable
in a character variable `group` taking the values `Basal`, `DRTA`, and `Strat` all
in the system file `c:/saslibrary/ex12.txt`. Then the program

```
data example;
infile 'c:/saslibrary/ex12.txt';
input group $ score;
cards;
proc glm data=example;
class group;
model score = group/clm;
means group/t alpha=.01;
output out=resan p=preds student=stdres;
run;
```

carries out a one-way ANOVA for the data in `score`, with the levels in `group`.
Figure 12.1 contains the ANOVA table, and we see that the P-value for the F
test of $H_0 : \mu_1 = \mu_2 = \mu_3$ versus $H_0 : \mu_1 \neq \mu_2$ or $\mu_1 \neq \mu_3$ is .3288, so H_0 is not
rejected. Also from this table the MSE gives the value $s^2 = 9.08658009$. By
specifying the **clm** option in the **model** statement the predicted values, the
residuals, and the 95% confidence intervals for the expected values for each
observation are printed, although we do not reproduce this listing (there are
66 observations) here. In this context the predicted value for each observation
in a cell is the mean of that cell, so all such observations have the same
predicted values. Similarly, the 95% confidence intervals for the expected
observations are the same for all observations in the same cell. In this case
we get the 95% confidence intervals

$$(9.21572394, 11.78427606)$$
$$(8.44299666, 11.01154879)$$
$$(7.85208757, 10.42063970)$$

for `Basal`, `DRTA` and `Strat` respectively.

In the **means** statement, we have asked for the means of `score` for each
value of the `group` variable. The **t** option asks that all three pairwise tests of

```
                           General Linear Models Procedure
Dependent Variable: SCORE

Source                DF         Sum of Squares       Mean Square   F Value    Pr > F
Model                  2            20.57575758        10.28787879     1.13     0.3288
Error                 63           572.45454545         9.08658009
Corrected Total       65           593.03030303

               R-Square                C.V.             Root MSE          SCORE Mean
               0.034696              30.79723          3.0143954          9.78787879

Source                DF             Type I SS         Mean Square   F Value    Pr > F
GROUP                  2            20.57575758        10.28787879     1.13     0.3288
Source                DF           Type III SS         Mean Square   F Value    Pr > F
GROUP                  2            20.57575758        10.28787879     1.13     0.3288
```

Figure 12.1: Output from **proc glm** for Example 12.6 in IPS.

equality of means be carried out using the two sample t procedure; i.e. using the statistics

$$t_{ij} = \frac{\bar{x}_i - \bar{x}_j}{s\sqrt{\frac{1}{n_i} + \frac{1}{n_j}}}$$

to test for equality of means between two groups, with a difference being declared if the *P*-value is smaller than `alpha=.01`. The output from the **means** statement is given in Figure 12.2, and we see that no differences are found.

The general form of the means statement is

means *effects/options*;

where *effects* specify the classifications for which the means are to be calculated for the response variable in the **model** statement. The *options* specify which *multiple comparison procedure* is to be used for the comparisons of the means. By specifying the option **t**, the *Fisher's LSD (least significant difference)* method is selected. Lowering the critical level is standard practice to make sure that, when conducting multiple tests of significance, the *family error rate* — the probability of declaring at least one result significant when no null hypotheses are false — is not too high. The value of α is referred to as the *individual error rate*. The default value of α, if the **alpha** option is not given, is .05. Many other multiple comparison procedures are available within **proc glm**, e.g., Bonferroni t tests (**bon**), Duncan's multiple-range

```
                 General Linear Models Procedure

                 T tests (LSD) for variable: SCORE

     NOTE: This test controls the type I comparisonwise error rate not the
           experimentwise error rate.

                 Alpha= 0.01  df= 63  MSE= 9.08658
                        Critical Value of T= 2.66
                 Least Significant Difference= 2.4141

     Means with the same letter are not significantly different.

           T Grouping              Mean      N   GROUP

                    A            10.5000     22   Basal
                    A
                    A             9.7273     22   DRTA
                    A
                    A             9.1364     22   Strat
```

Figure 12.2: Output from the **means** statement in **proc glm** for Example 12.6 in IPS.

test (**duncan**), Tukey's studentized range test (**tukey**), and Scheffé's multiple comparison procedure (**scheffe**); see reference 3 in Appendix E for a discussion of these and others.

Of course, it is important to carry out a residual analysis to check that the assumptions we have made are reasonable. So in the **output** statement of the above example we create a new data set `resan` containing the predicted values and the standardized residuals in the variables `preds` and `stdres`, respectively. Then the statements

```
axis1 length=6 in;
axis2 length=4 in;
symbol value=plus color=black;
symbol2 value=dot interpol=join color=black;
proc gplot data=resan;
plot stdres*group=1 preds*group=2/ haxis=axis1 vaxis=axis2;
```

give a scatterplot of the standardized residuals for each value of **group** in Figure 12.3 and plot the cell means for each group in Figure 12.4. The scatterplot for the standardized residuals indicates that the assumptions we have made seem quite reasonable for this data set — the scatters are roughly symmetrical about 0 and lie within $(-3, 3)$. Recall that we can get boxplots and normal probability plots for the standardized residuals as well using **proc univariate**. For evocative purposes, in the plot of Figure 12.4, we join the

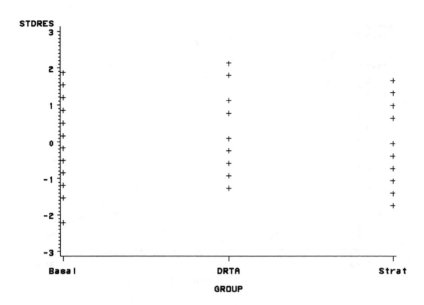

Figure 12.3: Plot of standardized residuals for fit of one-way ANOVA model in Example 12.6 of IPS.

cell means via lines. This plot seems to indicate some differences among the means although we have no justification for saying this from the results of the statistical tests we conducted.

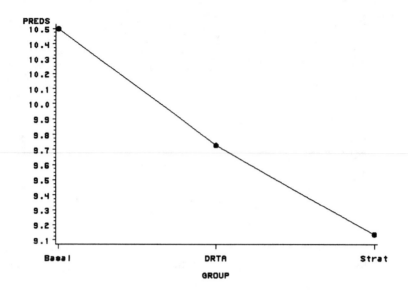

Figure 12.4: Plot of cell means for each value of **group** in Example 12.6 of IPS.

12.2 Exercises

When the data for an exercise come from an exercise in IPS, the IPS exercise number is given in parentheses ( ). All computations in these exercises are to be carried out using SAS and the exercises are designed to ensure that you have a reasonable understanding of the SAS material in this chapter. More generally you should be using SAS to do all the computations and plotting required for the problems in IPS.

1. Generate a sample of 10 from each of the $N(\mu_i, \sigma)$ distributions for $i = 1, \ldots, 5$, where $\mu_1 = 1, \mu_2 = 1, \mu_3 = 1, \mu_4 = 1, \mu_5 = 2$, and $\sigma = 3$. Carry out a one-way ANOVA and plot a normal quantile plot of the residuals and the residuals against the explanatory variable. Compute .95 confidence intervals for the means. Carry out Fisher's LSD procedure with the individual error rate set at .03.

2. Generate a sample of 10 from each of the $N(\mu_i, \sigma_i)$ distributions for $i = 1, \ldots, 5$, where $\mu_1 = 1, \mu_2 = 1, \mu_3 = 1, \mu_4 = 1, \mu_5 = 2$, and $\sigma_1 = \sigma_2 = \sigma_3 = \sigma_4 = 3$ and $\sigma_5 = 8$. Carry out a one-way ANOVA and

plot a normal quantile plot of the residuals and the residuals against the explanatory variable. Compare the residual plots with those obtained in Exercise 1.

3. If X_1 is distributed $Chisquare(k_1)$ independently of X_2, which is distributed $N(\delta, 1)$, then the random variable $Y = X_1 + X_2^2$ is distributed according to a *noncentral* $Chisquare(k + 1)$ distribution with noncentrality $\lambda = \delta^2$. Generate samples of $n = 1000$ from this distribution with $k = 2$ and $\lambda = 0, 1, 5, 10$. Plot histograms of these samples with the midpoints $0, 1, \dots, 200$. Comment on the appearance of the histograms.

4. If X_1 is distributed *noncentral* $Chisquare(k_1)$ with non-centrality λ independently of X_2, which is distributed $Chisquare(k_2)$, then the random variable

$$Y = \frac{X_1/k_1}{X_2/k_2}$$

is distributed according to a *noncentral* $F(k_1, k_2)$ distribution with noncentrality λ. Generate samples of $n = 1000$ from this distribution with $k_1 = 2, k_2 = 3$, and $\lambda = 0, 1, 5, 10$. Plot histograms of these samples with the midpoints $0, 1, \dots, 200$. Comment on the appearance of the histograms.

5. As noted in IPS, the F statistic in a one-way ANOVA, when the standard deviation σ is constant from one level to another, is distributed *noncentral* $F(k_1, k_2)$ with noncentrality λ, where $k_1 = I - 1$, $k_2 = n_1 + \cdots n_I - I$,

$$\lambda = \frac{\sum_{i=1}^{I} n_i \left(\mu_i - \bar{\mu}\right)^2}{\sigma^2}$$

and

$$\bar{\mu} = \frac{\sum_{i=1}^{I} n_i \mu_i}{\sum_{i=1}^{I} n_i}$$

Using simulation approximate the power of the test in Exercise 1 with level .05 and the values of the parameters specified by the estimates.

Chapter 13

Two-way Analysis of Variance

This chapter deals with methods for making inferences about the relationship existing between a single numeric response variable and two categorical explanatory variables. The **proc glm** procedure is used to carry out a two-way ANOVA.

We write the two-way ANOVA model as $x_{ijk} = \mu_{ij} + \epsilon_{ijk}$, where $i = 1, \ldots, I$ and $j = 1, \ldots, J$ index the levels of the categorical explanatory variables and $k = 1, \ldots, n_{ij}$ indexes the individual observations at each treatment (combination of levels), μ_{ij} is the mean response at the i-th level and the j-th level of the first and second explanatory variable, respectively, and the errors ϵ_{ijk} are a sample from the $N(0, \sigma)$ distribution. Based on the observed x_{ijk}, we want to make inferences about the unknown values of the parameters $\mu_{11}, \ldots, \mu_{IJ}, \sigma$.

13.1 Example

We consider a generated example where $I = J = 2$, $\mu_{11} = \mu_{21} = \mu_{12} = 1$, $\mu_{22} = 3$, $\sigma = 3$ and $n_{11} = n_{21} = n_{12} = n_{22} = 5$. The ϵ_{ijk} are generated as a sample from the $N(0, \sigma)$ distribution, and then we put $x_{ijk} = \mu_{ij} + \epsilon_{ijk}$ for $i = 1, \ldots, I$ and $j = 1, \ldots, J$ and $k = 1, \ldots, n_{ij}$. Then we pretend that we don't know the values of the parameters and carry out a two-way analysis of variance. The x_{ijk} are stored in the variable **respons**, the values of i in

factor1, and the values of j in factor2, all in the SAS data set example.
We generate the SAS data set using the program

```
data example;
seed=845443;
mu11=1;
mu21=1;
mu12=1;
mu22=3;
sigma=3;
do i=1 to 2;
do j=1 to 2;
do k=1 to 5;
factor1=i;
factor2=j;
if i=1 and j=1 then
x=mu11+sigma*rannor(seed);
if i=1 and j=1 then
factor='11';
if i=2 and j=1 then
x=mu21+sigma*rannor(seed);
if i=2 and j=1 then
factor='21';
if i=1 and j=2 then
x=mu12+sigma*rannor(seed);
if i=1 and j=2 then
factor='12';
if i=2 and j=2 then
x=mu22+sigma*rannor(seed);
if i=2 and j=2 then
factor='22';
output example;
end;
end;
end;
```

which is a bit clumsy, and we note that we can do this more efficiently using
arrays, discussed in Appendix B, or **proc iml**, discussed in Appendix C.
Note that we have created a character variable factor that is equal to one

of the four values 11, 21, 12, 22. This proves useful when we want to plot the data or residuals in side-by-side scatterplots as we will see. In any case, given that we have generated the data in **example** as described, we proceed to carry out a two-way ANOVA using

```
proc glm data=example;
class factor1 factor2;
model x = factor1 factor2 factor1*factor2;
means factor1 factor2 factor1*factor2;
output out=resan p=preds student=stdres;
```

which produces the output shown in Figures 13.1 and 13.2. We see that the P-value for testing H_0 : no interaction is given by the entry in the row labeled **FACTOR1*FACTOR2** and equals .0165, so we reject this null hypothesis. Now there is no point in continuing to test for an effect due to **factor1** or an effect due to **factor2** (using the rows labeled **FACTOR1** and **FACTOR2**, respectively) because we know that both variables must have an effect if there is an interaction. To analyze just what the interaction effect is we look at the output from the **means** statement given in Figure 13.2. In the **means** statement, we asked for the cell means to be printed for each level of **factor1**, each level of **factor2**, and each value of (**factor1, factor2**). When there is an interaction, it is the cell means for (**factor1, factor2**) that we must look at to determine just what effect the explanatory variables are having. We do this by plotting them and by carrying two-sample t tests comparing the means. Note that while the multiple comparison procedures are still available as options to the **means** statement, they work only with main effects: comparisons of the means for levels of individual variables. SAS does not permit the use of multiple comparison procedures with interaction effects to discourage specifying too many tests.

Of course we must also check our assumptions to determine the validity of the analysis. We note that we saved the predicted values and the standardized residuals in the SAS data set **resan** so that these variables would be available for a residual analysis. The statements

```
axis1 length=6 in;
axis2 length=4 in;
symbol interpol=box;
proc gplot data=resan;
plot stdres*factor=1/ haxis=axis1 vaxis=axis2;
```

```
                        General Linear Models Procedure
Dependent Variable: X

Source                 DF           Sum of Squares      Mean Square   F Value   Pr > F
Model                   3           144.44701292        48.14900431    4.90     0.0133
Error                  16           157.34529536         9.83408096
Corrected Total        19           301.79230828

              R-Square                    C.V.              Root MSE            X Mean
              0.478631                  145.1059           3.1359338         2.16113426

Source                 DF              Type I SS         Mean Square   F Value   Pr > F
FACTOR1                 1             46.96759225        46.96759225    4.78     0.0441
FACTOR2                 1             27.03196972        27.03196972    2.75     0.1168
FACTOR1*FACTOR2         1             70.44745094        70.44745094    7.16     0.0165

Source                 DF              Type III SS       Mean Square   F Value   Pr > F
FACTOR1                 1             46.96759225        46.96759225    4.78     0.0441
FACTOR2                 1             27.03196972        27.03196972    2.75     0.1168
FACTOR1*FACTOR2         1             70.44745094        70.44745094    7.16     0.0165
```

Figure 13.1: Two-way ANOVA table from **proc glm**.

```
               General Linear Models Procedure

   Level of             --------------X--------------
   FACTOR1      N            Mean              SD

     1          10        0.62869189        2.46603546
     2          10        3.69357663        4.71513801

   Level of             --------------X--------------
   FACTOR2      N            Mean              SD

     1          10        0.99855158        1.16837415
     2          10        3.32371694        5.40035447

 Level of    Level of           --------------X--------------
 FACTOR1     FACTOR2     N           Mean              SD

    1           1        5        1.34290770        0.94348748
    1           2        5       -0.08552391        3.39375388
    2           1        5        0.65419547        1.37289816
    2           2        5        6.73295779        5.00437213
```

Figure 13.2: Output from **means** statement in **proc glm**.

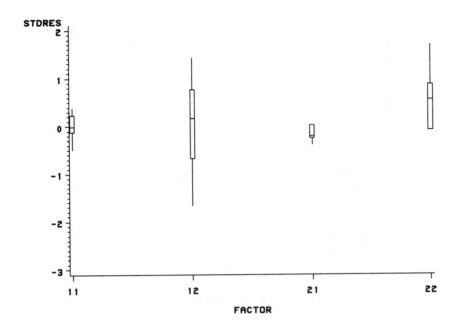

Figure 13.3: Side-by-side boxplots of standardized residuals for each cell.

result in the side-by-side boxplots of the standardized residuals given in Figure 13.3. The residuals don't appear to have the same variance — even though we know that all the assumptions hold in this example. On the other hand, the sample sizes are small, so we can expect some plots like this to look wrong. Of course other plots of the residuals should also be looked at.

To examine the interaction effect we first looked at a plot of the cell means. The statements

```
axis1 length=6 in;
axis2 length=4 in;
symbol value=dot interpol=join;
proc gplot data=resan;
plot preds*factor1=factor2/ haxis=axis1 vaxis=axis2;
```

produce the plots of the response curves shown in Figure 13.4; the means are plotted and joined by lines for each level of `factor2`. This gives an indication of the form of the relationship. In particular, as `factor1` goes from level 1 to level 2, the mean response increases when `factor2=2` but stays about the

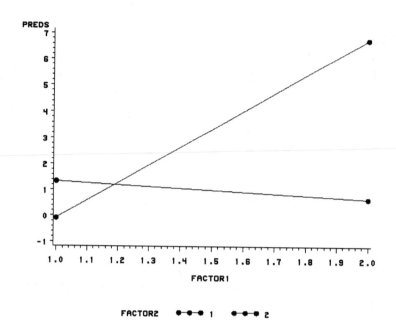

Figure 13.4: Plot of response curves.

same when `factor2=1`, and, of course, this is correct. The statements

```
data comp;
set example;
if factor2=2;
proc ttest data=comp;
class factor1;
var x;
```

first construct the SAS data set `comp` by selecting the observations from `example` where `factor2=2` and then uses **proc ttest** to test for a difference between the means when `factor1=1` and `factor1=2`. The output in Figure 13.5 indicates that we would reject the null hypothesis of no difference between these means. Other comparisons of means can be carried out in a similar fashion.

It is possible to also fit a two-way model without an interaction effect. In the example could do this with the **model** statement

```
model x = factor1 factor2;
```

```
                                    TTEST PROCEDURE

Variable: X
FACTOR1      N        Mean      Std Dev    Std Error   Variances        T      DF    Prob>|T|
----------------------------------------------------    ----------------------------------------
   1         5    -0.08552391  3.39375388  1.51773287   Unequal    -2.5215    7.0    0.0396
   2         5     6.73295779  5.00437213  2.23802326   Equal      -2.5215    8.0    0.0357

For H0: Variances are equal, F' = 2.17   DF = (4,4)    Prob>F' = 0.4704
```

Figure 13.5: Output from **proc ttest** comparing the means when `factor1=1`, `factor2=2` and `factor1=2`, `factor2=2`.

and the ANOVA table would include the sums of squares corresponding to interaction with the error. This would give more degrees of freedom for estimating error and so could be regarded as a good thing. On the other hand, it is worth noting that we would in addition be assuming no interaction so we should play close attention to our residual analysis to make sure that the assumption is warranted.

13.2   Exercises

When the data for an exercise come from an exercise in IPS, the IPS exercise number is given in parentheses ( ). All computations in these exercises are to be carried out using SAS, and the exercises are designed to reinforce your understanding of the SAS material in this chapter. More generally, you should use SAS to do all the computations and plotting required for the problems in IPS.

1. Suppose $I = J = 2$, $\mu_{11} = \mu_{21} = 1$ and $\mu_{12} = \mu_{22} = 2$, and $\sigma = 2$, and $n_{11} = n_{21} = n_{12} = n_{22} = 10$. Generate the data for this situation and carry out a two-way analysis. Plot the cell means (an interaction effect plot). Do your conclusions agree with what you know to be true?

2. Suppose $I = J = 2$, $\mu_{11} = \mu_{21} = 1$, and $\mu_{12} = 3$, $\mu_{22} = 2$, and $\sigma = 2$, and $n_{11} = n_{21} = n_{12} = n_{22} = 10$. Generate the data for this situation and carry out a two-way analysis. Plot the cell means (an interaction effect plot). Do your conclusions agree with what you know to be true?

3. Suppose $I = J = 2$, $\mu_{11} = \mu_{21} = 1$, and $\mu_{12} = \mu_{22} = 2$, and $\sigma = 2$ and $n_{11} = n_{21} = n_{12} = n_{22} = 10$. Generate the data for this situation

and carry out a two-way analysis. Form 95% confidence intervals for the marginal means. Repeat your analysis using the model without an interaction effect and compare the confidence intervals. Can you explain your results?

Chapter 14

Nonparametric Tests

SAS statement introduced in this chapter
proc npar1way

This chapter deals with inference methods that do not depend on the assumption of normality. These methods are sometimes called *nonparametric* or *distribution-free* methods. Recall that we discussed a distribution-free method in Section II.7.3 where we presented the sign test for the median.

14.1   PROC NPAR1WAY

This procedure provides for *nonparametric analyses* for testing that a random variable has the same distribution across different groups. The procedure analyzes only one-way classifications. The tests are all asymptotic and hence require large sample sizes for validity.

Following are some of the statements are available with **proc npar1way**.

proc npar1way *options*;
var *variables*;
class *variable*;
by *variables*;

The **class** statement must appear. Note that only one classification variable may be specified. The **var** *variables*; statement identifies response variables

189

for which we want an analysis performed. The **by** statement works as described in **proc sort**.

Following are some of the options available with the **proc npar1way** statement.

data = *SASdataset*
wilcoxon

We describe the use of **wilcoxon** in Sections II.14.2 and II.14.4.

Note that the analysis techniques used by **proc npar1way** assume that the distribution form is the same in each classification group but may differ in their locations only.

14.2   Wilcoxon Rank Sum Test

The Wilcoxon rank sum procedure tests for a difference in the medians of two distributions that differ at most in their medians. If $y_{11}, \ldots, y_{1n_1}$ is a sample from a continuous distribution with median ζ_1 and $y_{21}, \ldots, y_{2n_2}$ is a sample from a continuous distribution with median ζ_2, to test $H_0 : \zeta_1 = \zeta_2$ versus one of the alternatives $H_a : \zeta_1 \neq \zeta_2, H_a : \zeta_1 < \zeta_2$ or $H_a : \zeta_1 > \zeta_2$, we use the Wilcoxon rank sum test, available via the **wilcoxon** option in the **proc npar1way** statement.

For Example 14.1 of IPS we store the values of the class variable in `weed` and the responses in the variable `yield`. Then the program

```
data example;
input weeds yield;
cards;
0 166.7
0 172.2
0 165
0 176.9
3 158.6
3 176.4
3 153.1
3 156
proc npar1way wilcoxon data=example;
class weeds;
var yield;
```

```
                 N P A R 1 W A Y   P R O C E D U R E

           Wilcoxon Scores (Rank Sums) for Variable YIELD
                     Classified by Variable WEEDS

                        Sum of      Expected       Std Dev           Mean
     WEEDS       N      Scores      Under H0       Under H0          Score

       0         4       23.0         18.0       3.46410162      5.75000000
       3         4       13.0         18.0       3.46410162      3.25000000

       Wilcoxon 2-Sample Test (Normal Approximation)
       (with Continuity Correction of .5)

       S =   23.0000        Z =  1.29904        Prob > |Z| = 0.1939

       T-Test Approx. Significance = 0.2351

       Kruskal-Wallis Test (Chi-Square Approximation)
       CHISQ =   2.0833        DF = 1         Prob > CHISQ = 0.1489
```

Figure 14.1: Output from **proc npar1way** with **wilcoxon** option.

produces the output shown in Figure 14.1. This gives the value of the Wilcoxon rank sum test statistic as S=23 and the *approximate* P-value for testing $H_0 : \zeta_1 = \zeta_2$ versus $H_a : \zeta_1 \neq \zeta_2$ as Prob $> |Z|$ = .1939 where Z=1.22904,. Let us denote this P-value by P_2 to indicate that it arises from testing H_0 against the two-sided alternative. If instead we want to test H_0 against the one-sided alternative $H_a : \zeta_1 > \zeta_2$, the approximate P-value is given by $.5P_2$ when Z > 0, $1 - .5P_2$ when Z < 0. In this case the approximate P-value for testing $H_0 : \zeta_1 = \zeta_2$ versus $H_a : \zeta_1 > \zeta_2$ is given by $.5P_2 = .5(.1939) = 0.09695$, so we don't reject H_0. If we want to test H_0 against the one-sided alternative $H_a : \zeta_1 < \zeta_2$, the approximate P-value is given by $.5P_2$ when Z < 0, $1 - .5P_2$ when Z > 0.

The Wilcoxon rank sum test for a difference between the locations of two distributions is equivalent to another nonparametric test called the Mann-Whitney test. Suppose we have two independent samples $y_{11}, \ldots, y_{1n_1}$ and $y_{21}, \ldots, y_{2n_2}$ from two distributions that differ at most in their locations as represented by their medians; in other words one distribution can be obtained from the other by a location change. The Mann-Whitney statistic U is the number of pairs (y_{1i}, y_{2j}) where $y_{1i} > y_{2j}$ while the Wilcoxon rank sum test statistic W is the sum of the ranks from the first sample when the ranks are computed for the two samples considered as one sample. Then it can be shown that $W = U + n_1(n_{1+1})/2$.

14.3   Sign Test and Wilcoxon Signed Rank Test

We denote the median of a continuous distribution by ζ. As already mentioned in Section II.7.3, the value M(Sign), which equals the sign test statistic minus its mean under $H_0 : \zeta = 0$, and the P-value for testing H_0 against $H_a : \zeta \neq 0$ are printed out by **proc univariate**. We denote this P-value by P_2. Suppose we want to instead test $H_0 : \zeta = 0$ versus $H_a : \zeta > 0$. Then if M(Sign)> 0, the relevant P-value is $.5P_2$, and if M(Sign)< 0, the relevant P-value is $1 - .5P_2$. Suppose we want to test $H_0 : \zeta = 0$ versus $H_a : \zeta < 0$. Then if M(Sign)< 0, the relevant P-value is $.5P_2$, and if M(Sign)> 0, the relevant P-value is $1 - .5P_2$. We can also use the sign test when we have paired samples to test that the median of the distribution of the difference between the two measurements on each individual is 0. For this we just apply **proc univariate** to the differences.

The Wilcoxon signed rank test can also be used for inferences about the median of a distribution. The Wilcoxon signed rank test is based on the ranks of sample values, which is not the case for the sign test. The Wilcoxon signed rank test for the median also differs from the sign test in that it requires an assumption that the response values come from a continuous distribution symmetric about its median, while the sign test requires only a continuous distribution. The procedure **proc univariate** also prints out Sgn Rank, which is the value of the Wilcoxon signed rank statistic minus its mean when H_0 is true, and the value P_2, which is the P-value for testing $H_0 : \zeta = 0$ versus $H_a : \zeta \neq 0$. If instead we wish to test a one-sided hypothesis, then we can compute the P-value using the values of Sgn Rank and P_2 as discussed for the sign test.

Consider the data of Example 14.8 in IPS, where the differences between two scores is input as the variable diff in the SAS data set example, and suppose we want to test $H_0 : \zeta = 0$ versus $H_a : \zeta > 0$. Then the program

```
data example;
input diff;
cards;
.37
-.23
.66
-.08
```

```
M(Sign)          -0.5   Pr>=|M|      1.0000
Sgn Rank          1.5   Pr>=|S|      0.8125
```

Figure 14.2: Part of the output from **proc univariate** applied to the variable `diff` for Example 14.8 of IPS.

```
-.17
proc univariate data=example;
var diff;
run;
```

produces as part of its output the value of the sign test statistic and the value of the Wilcoxon signed rank statistic and their associated *P*-values, are shown in Figure 14.2. We see that the value of **Sgn Rank** is 1.5 and P_2 is .8125. The appropriate *P*-value is .5(.8125) = .40625, so we would not reject H_0.

14.4 Kruskal-Wallis Test

The Kruskal-Wallis test is the analog of the one-way ANOVA in the nonparametric setting. The distributions being compared are assumed to differ at most in their medians. This test can be carried out in SAS by using **proc nonpar1way** with the **wilcoxon** option. Suppose the data for Example 14.13 in IPS are stored in the SAS data set **corn** with weeds per meter in the variable **weeds** and corn yield in the variable **yield**. Then the program

```
data example;
input weeds yield;
cards;
0 166.7
0 172.2
0 165.0
0 176.9
1 166.2
1 157.3
1 166.7
1 161.1
```

```
                    N P A R I W A Y   P R O C E D U R E

               Wilcoxon Scores (Rank Sums) for Variable YIELD
                       Classified by Variable WEEDS

                              Sum of      Expected      Std Dev         Mean
    WEEDS         N           Scores      Under H0      Under H0        Score
      0           4        52.5000000        34.0       8.24014563   13.1250000
      1           4        33.5000000        34.0       8.24014563    8.3750000
      3           4        25.0000000        34.0       8.24014563    6.2500000
      9           4        25.0000000        34.0       8.24014563    6.2500000
                        Average Scores Were Used for Ties

            Kruskal-Wallis Test (Chi-Square Approximation)
            CHISQ =  5.5725        DF =  3        Prob > CHISQ = 0.1344
```

Figure 14.3: Output from **proc nonpar1way** with **wilcoxon** option with more than two groups.

```
3  158.6
3  176.4
3  153.1
3  156.0
9  162.8
9  142.4
9  162.7
9  162.4
proc npar1way wilcoxon data=example;
class weeds;
var yield;
run;
```

produces the output shown in Figure 14.3. We see a P-value of .1344 for testing H_0 : each sample comes from the same distribution versus H_a : at least two of the samples come from different distributions. Accordingly, we do not reject H_0.

14.5   Exercises

When the data for an exercise come from an exercise in IPS, the IPS exercise number is given in parentheses ( ). All computations in these exercises are to be carried out using SAS, and the exercises are designed to reinforce your understanding of the SAS material in this chapter. More generally, you should use SAS to do all the computations and plotting required for the problems in IPS.

1. Generate a sample of $n = 10$ from the $N(0, 1)$ distribution and compute the P-value for testing H_0 : the median is 0 versus H_a : the median is not 0, using the t test and the Wilcoxon signed rank test. Compare the P-values. Repeat this exercise with $n = 100$.

2. Generate two samples of $n = 10$ from the $Student(1)$ distribution and to the second sample add 1. Then test H_0 : the medians of the two distributions are identical versus H_a : the medians are not equal using the two sample t test and the Wilcoxon rank sum test. Compare the results.

3. Generate a sample of 10 from each of the $N(1, 2)$, $N(2, 2)$, and $N(3, 1)$ distributions. Test for a difference among the distributions using a one-way ANOVA and the Kruskal-Wallis test. Compare the results.

Chapter 15

Logistic Regression

SAS statement introduced in this chapter

proc logistic

This chapter deals with the *logistic regression model*. This model arises when the response variable y is binary, i.e., takes only two values, and we have a number of explanatory variables $x_1, \ldots, x_k$. In SAS we use **proc logistic** for carrying out logistic regression.

15.1   Logistic Regression Model

The regression techniques discussed in Chapters 10 and 11 of IPS require the response variable y to be a continuous variable. In many contexts, however, the response is discrete and in fact binary. It takes the values 0 and 1. Let p denote the probability of a 1. This probability is related to the values of the explanatory variables $x_1, \ldots, x_k$. We cannot, however, write this as $p = \beta_0 + \beta_1 x_1 + \ldots + \beta_k x_k$ because the right-hand side is not constrained to lie in the interval $[0, 1]$, which it must if it is to represent a probability. One solution to this problem is to employ the *logit link function,* given by

$$\ln\left(\frac{p}{1-p}\right) = \beta_0 + \beta_1 x_1 + \cdots + \beta_k x_k$$

and this leads to the equations

$$\frac{p}{1-p} = \exp\left\{\beta_0 + \beta_1 x_1 + \cdots + \beta_k x_k\right\}$$

and

$$p = \frac{\exp\left\{\beta_0 + \beta_1 x_1 + \cdots + \beta_k x_k\right\}}{1 + \exp\left\{\beta_0 + \beta_1 x_1 + \cdots + \beta_k x_k\right\}}$$

for the *odds* $p/(1-p)$ and probability p, respectively. The right-hand side of the equation for p is now always between 0 and 1. Note that logistic regression is based on an ordinary regression relation between the logarithm of the odds in favor of the event occurring at a particular setting of the explanatory variables and the values of the explanatory variables $x_1, \ldots, x_k$. The quantity $\ln\left(p/(1-p)\right)$ is referred to as the *log odds*.

The procedure for estimating the coefficients $\beta_0, \beta_1, \ldots, \beta_k$ using this relation and carrying out tests of significance on these values is known as *logistic regression*. Typically, more sophisticated statistical methods than least squares are needed for fitting and inference in this context, and we rely on software such as SAS to carry out the necessary computations.

Other link functions that are often used are available in SAS. In particular the *probit link function* is given by

$$\Phi^{-1}(p) = \beta_0 + \beta_1 x_1 + \cdots + \beta_k x_k$$

where Φ is the cumulative distribution function of the $N(0,1)$ distribution, and this leads to the relation

$$p = \Phi\left(\beta_0 + \beta_1 x_1 + \cdots + \beta_k x_k\right)$$

which is also always between 0 and 1. We restrict our attention here to the logit link function.

15.2   Example

Suppose we have the following 10 observations

	y	x1	x2
1	0	-0.65917	0.43450
2	0	0.69408	0.48175
3	1	-0.28772	0.08279
4	1	0.76911	0.59153
5	1	1.44037	2.07466
6	0	0.52674	0.27745
7	1	0.38593	0.14894
8	1	-0.00027	0.00000
9	0	1.15681	1.33822
10	1	0.60793	0.36958

where the response is y and the predictor variables are x_1 and x_2 (note that $x_2 = x_1^2$). Suppose we want to fit the model

$$\ln\left(\frac{p}{1-p}\right) = \beta_0 + \beta_1 x_1 + \beta_2 x_2$$

and conduct statistical inference concerning the parameters of the model. Then the program

```
data example;
input y x1 x2;
cards;
0 -0.65917 0.43450
0  0.69408 0.48175
1 -0.28772 0.08279
1  0.76911 0.59153
1  1.44037 2.07466
0  0.52674 0.27745
1  0.38593 0.14894
1 -0.00027 0.00000
0  1.15681 1.33822
1  0.60793 0.36958
proc logistic;
model y = x1 x2;
run;
```

fits the model and computes various test statistics, as given in Figure 15.1. The fitted model is given by

$$\ln\left(\frac{p}{1-p}\right) = -0.5228 - 0.7400x_1 + 0.7796x_2$$

and the standard errors of the estimates of β_0, β_1, and β_2 are recorded as 0.9031, 1.6050, and 1.5844, respectively, so we can see that these quantities are not being estimated with great accuracy. The output also gives the P-value for $H_0 : \beta_0 = 0$ versus $H_a : \beta_0 \neq 0$ as 0.5627, the P-value for $H_0 : \beta_1 = 0$ versus $H_a : \beta_1 \neq 0$ as 0.6448, and the P-value for $H_0 : \beta_2 = 0$ versus $H_a : \beta_2 \neq 0$ as 0.6227. Further the test of $H_0 : \beta_1 = \beta_2 = 0$ versus $H_a : \beta_1 \neq 0$ or $\beta_2 \neq 0$ is recorded as -2 LOG L and has P-value .8759. In this example there is no evidence of any nonzero coefficients. Note that when $\beta_0 = \beta_1 = \beta_2 = 0$, $p = .5$.

Also provided in the output is the estimate 0.477 for the odds ratio for x_1. The odds ratio for x_1 is given by $\exp\left(\beta_1\right)$, which is the ratio of the odds at $x_1 + 1$ to the odds at x_1 when x_2 is held fixed or when $\beta_2 = 0$. Since there is evidence that $\beta_2 = 0$ (P-value $= .6227$), the odds ratio has a direct interpretation. Note, however, that if this wasn't the case then the odds ratio would not have such an interpretation; it doesn't makes sense for x_2 to be held fixed when x_1 changes in this example because they are functionally related variables. Similar comments apply to the estimate 2.181 for the odds ratio for x_2.

Often the data come to us in the form of counts; namely, for each setting of the explanatory variables we get the value (r, n), where r corresponds to the number of trials — e.g., tosses of a coin — and n corresponds to the number of events — e.g., heads — that occurred. So n is $Binomial\left(r, p\right)$ distributed with p dependent on the values of the explanatory variables through the logistic link. For example, suppose we have two explanatory variables x_1 and x_2 taking the values $(-1, -1), (-1, 1), (1, -1), (1, 1)$ and we observe the values $(20, 13), (15, 10), (18, 11), (20, 12)$ respectively for (r, n). Then the program

```
data example;
input r n x1 x2;
cards;
20 13 -1 -1
15 10 -1 1
```

```
                    Response Profile

                Ordered
                 Value       Y        Count

                   1         0          4
                   2         1          6


      Model Fitting Information and Testing Global Null Hypothesis BETA=0

                                    Intercept
                         Intercept     and
          Criterion        Only     Covariates    Chi-Square for Covariates

          AIC             15.460      19.195            .
          SC              15.763      20.103            .
          -2 LOG L        13.460      13.195        0.265 with 2 DF (p=0.8759)
          Score             .           .          0.268 with 2 DF (p=0.8747)


                    Analysis of Maximum Likelihood Estimates

                Parameter  Standard      Wald        Pr >      Standardized    Odds
Variable   DF   Estimate     Error    Chi-Square  Chi-Square    Estimate      Ratio

INTERCPT    1    -0.5228    0.9031      0.3351      0.5627          .            .
X1          1    -0.7400    1.6050      0.2125      0.6448      -0.259879      0.477
X2          1     0.7796    1.5844      0.2421      0.6227       0.277269      2.181


          Association of Predicted Probabilities and Observed Responses

              Concordant = 50.0%      Somers' D = 0.042
              Discordant = 45.8%      Gamma     = 0.043
              Tied       =  4.2%      Tau-a     = 0.022
              (24 pairs)              c         = 0.521
```

Figure 15.1: Output from **proc logistic**.

```
18 11 1 -1
20 12 1 1
proc logistic;
model n/r= x1 x2;
run;
```

produces the output like that shown in Figure 15.1; we fit the model

$$\ln\left(\frac{p}{1-p}\right) = \beta_0 + \beta_1 x_1 + \beta_2 x_2$$

and carry out various tests of significance. Note the form of the **model** statement in this case. Many other aspects of fitting logistic regression models are available in SAS, and we refer the reader to reference 4 in Appendix E for a discussion of these.

15.3   Exercises

When the data for an exercise come from an exercise in IPS, the IPS exercise number is given in parentheses ( ). All computations in these exercises are to be carried out using SAS, and the exercises are designed to reinforce your understanding of the SAS material in this chapter. More generally, you should use SAS to do all the computations and plotting required for the problems in IPS.

1. Generate a sample of 20 from the *Bernoulli*(.25) distribution. Pretending that we don't know p compute a 95% confidence interval for this quantity. Using this confidence interval, form 95% confidence intervals for the odds and the log odds.

2. Let x take the values -1, $-.5$, 0, $.5$ and 1. Plot the log odds

$$\ln\left(\frac{p}{1-p}\right) = \beta_0 + \beta_1 x$$

against x when $\beta_0 = 1$ and $\beta_1 = 2$. Plot the odds and the probability p against x.

3. Let x take the values -1, $-.5$, 0, $.5$, and 1. At each of these values generate a sample of four values from the *Bernoulli*(p_x) distribution where

$$p_x = \frac{\exp\{1 + 2x\}}{1 + \exp\{1 + 2x\}}$$

and let these values be the y response values. Carry out a logistic regression analysis of these data using the model

$$\ln\left(\frac{p_x}{1-p_x}\right) = \beta_0 + \beta_1 x$$

Test the null hypothesis $H_0 : \beta_1 = 0$ versus $H_0 : \beta_1 \neq 0$ and determine if the correct inference was made.

4. Let x take the values -1, $-.5$, 0, $.5$, and 1. At each of these values generate a sample of four values from the *Bernoulli*(p_x) distribution where

$$p_x = \frac{\exp\{1 + 2x\}}{1 + \exp\{1 + 2x\}}$$

and let these values be the y response values. Carry out a logistic regression analysis of these data using the model

$$\ln\left(\frac{p_x}{1 - p_x}\right) = \beta_0 + \beta_1 x + \beta_2 x^2$$

Test the null hypothesis $H_0 : \beta_2 = 0$ versus $H_a : \beta_2 \neq 0$.

5. Let x take the values -1, $-.5$, 0, $.5$, and 1. At each of these values, generate a sample of four values from the *Bernoulli*$(.5)$ distribution. Carry out a logistic regression analysis of these data using the model

$$\ln\left(\frac{p_x}{1 - p_x}\right) = \beta_0 + \beta_1 x + \beta_2 x^2$$

Test the null hypothesis $H_0 : \beta_1 = \beta_2 = 0$ versus $H_a : \beta_1 \neq 0$ or $\beta_2 \neq 0$.

Part III

Appendices

Appendix A

Operators and Functions in the Data Step

A.1   Operators

A number of different operators can be used in SAS programs. There is a priority for how expressions are evaluated from left to right, but we advocate the use of parentheses ( ) to make expressions easier to read. Here we group operators by type.

A.1.1   Arithmetic Operators

Arithmetic operators indicate that an arithmetic calculation is to be performed. The arithmetic operators are:

** exponentiation; e.g., x**y is "x raised to the power y, or x^y"

* multiplication

/ division

+ addition

− subtraction

If a missing value is an operand for an arithmetic operator, the result is a missing value.

A.1.2   Comparison Operators

Comparison operators propose a relationship between two quantities and ask SAS to determine whether or not that relationship holds. As such, the output from a comparison operation is 1 if the proposed comparison is true and 0 if the proposed comparison is false.

= or EQ equal to
NE not equal to
> or GT greater than
NG not greater than
< or LT less than
NL not less than
>= or GE greater than or equal to
<= or LE less than or equal to

A.1.3   Logical Operators

Logical operators, also called Boolean operators, are usually used in expressions to link sequences of comparisons. The logical operators are:

& and
| or
~ not

A.1.4   Other Operators

The operators in this category are:

>< minimum of two surrounding quantities
<> maximum of two surrounding quantities
|| concatenation of two character values

A.1.5   Priority of Operators

Expressions within parentheses are evaluated before those outside.

The operations **, + prefix, - prefix, ><, <> are then performed, followed by * and /, which is followed by + and -, then ||, then <, <=, =, >=, >, followed by &, and finally, |.

Operations with the same priority are performed in the order in which they appear.

When in doubt, use parentheses.

A.2 Functions

A SAS function is a routine that returns a value computed from arguments.

A.2.1 Arithmetic Functions

abs(x) returns the absolute value of x.

max(*arguments*) returns the largest value among the *arguments*. There may be two or more arguments separated by commas or a variable range list preceded by `of`. For example, `max(1,2,3)` is 3, `max(x,y,z)` gives the maximum of `x`, `y`, and `z`, and `max(of x1-x100)` gives the maximum of `x1`, `x2`, , `x100`.

min(*arguments*) returns the smallest value among the *arguments*. The *arguments* obey the same rules as with **max**.

mod(*argument1, argument2*) calculates the remainder when the quotient of *argument1* divided by *argument2* is calculated.

sign(x) returns a value of -1 if $x < 0$, a value of 0 if $x = 0$, a value of $+1$ if $x > 0$.

sqrt(x) calculates the square root of the value of x. The value of x must be nonnegative.

A.2.2 Truncation Functions

ceil(x) results in the smallest integer larger than x.

floor(x) results in the largest integer smaller than the argument.

int(x) results in the integer portion of the value of x.

round(*value, roundoffunit*) rounds a *value* to the nearest roundoff unit. The value of the *roundoffunit* must be greater than zero. If the *roundoffunit* is omitted, a value of 1 is used and *value* is rounded to the nearest integer.

A.2.3   Special Functions

digamma(x) computes the derivative of the log of the gamma function. The value of this function is undefined for nonpositive integers.

exp(x) raises $e = 2.71828$ to the power specified by x.

gamma(x) produces the complete gamma function. If x is an integer, then **gamma**(x) is $(x - 1)!$ (i.e. the factorial of $(x - 1)$).

lgamma(x) results in the natural logarithm of the **gamma** function of the value of x.

log(x) results in the natural logarithm (base e) of the value of x. The value of x must be a positive value.

log10(x) results in the common logarithm $(\log_{10})$ of the value of x.

trigamma(x) returns the derivative of the **digamma** function.

A.2.4   Trigonometric Functions

arcos(x) returns the inverse cosine of the value of x. The result is in radians and $-1 \leq x \leq +1$.

arsin(x) returns the inverse sine of the value of x. The result is in radians and $-1 \leq x \leq +1$.

atan(x) returns the inverse tangent of the value of x.

cos(x) returns the cosine of the value of x. The value of x is assumed to be in radians.

cosh(x) returns the hyperbolic cosine of the value of x.

sin(x) returns the sine of the value of x. The value of x is assumed to be in radians.

sinh(x) returns the hyperbolic sine of the value of x.

tan(x) returns the tangent of the value of x. The value of x is assumed to be in radians; it may not be an odd multiple of $\frac{\pi}{2}$.

tanh(x) returns the hyperbolic tangent of the value of x.

A.2.5   Probability Functions

betainv(p, a, b), where $0 \leq p \leq 1$, $a > 0$, and $b > 0$, returns the Beta(a, b) inverse distribution function at p.

cinv(p, df) returns the inverse distribution function of the Chisquare(df) distribution at p, $0 \leq p \leq 1$.

finv(p, ndf, ddf) returns the inverse distribution function for the F(ndf, ddf) distribution at p, $0 \leq p \leq 1$.

gaminv(p, α), where $0 < p < 1$ and $\alpha > 0$, computes the Gamma$(\alpha, 1)$ inverse distribution function at p.

poisson(λ, n), where $0 \leq \lambda$ and $0 \leq n$, returns the Poisson(λ) cdf at n.

probbeta(x, a, b), where $0 \leq x \leq 1$ and $0 < a, b$, returns the Beta(a, b) cdf at x.

probbnml(p, n, m), where $0 \leq p \leq 1$, $1 \leq n$, $0 \leq m \leq n$, returns the Binomial(n, p) cdf at m.

probchi(x, df) returns the Chisquare(df) cdf at x.

probf(x, ndf, ddf) returns the F(ndf, ddf) cdf at x.

probgam(x, α) returns the Gamma$(\alpha, 1)$ cdf at x.

probhypr(nn, k, n, x) returns the Hypergeometric(nn, k, n) cdf at x where $\max(0, k + n - nn) \leq x \leq \min(k, n)$.

probit(p) returns the $N(0, 1)$ inverse distribution function at p.

probnegb(p, n, m) where $0 \leq p \leq 1$, $1 \leq n$, $0 \leq m$, returns the Negative Binomial(n, p) cdf at m.

probnorm(x) returns the $N(0, 1)$ cdf at x.

probt(x, df) returns the Student(df) cdf at x.

tinv(p, df) returns the Student(df) inverse distribution function at p, $0 \leq p \leq 1$.

A.2.6 Sample Statistical Functions

In the following functions, *arguments* may be a list of numbers separated by commas, a list of variables separated by commas, or a variable range list preceded by `of`.

css(*arguments*) results in the corrected sum of squares of the *arguments*.

cv(*arguments*) results in the coefficient of variation of the *arguments*.

kurtosis(*arguments*) results in the kurtosis statistic of the *arguments*.

mean(*arguments*) results in the average of the values of the *arguments*.

n(*arguments*) returns the number of nonmissing *arguments*.

nmiss(*arguments*) gives the number of missing values in a string of *arguments*.

range(*arguments*) gives the range of the values of the *arguments*.

skewness(*arguments*) results in a measure of the skewness of the *arguments* values.

std(*arguments*) gives the standard deviation of the values of the *arguments*.

stderr(*arguments*) results in the standard error of the mean of the values of the *arguments*.

sum(*arguments*) results in the sum of the *arguments*.

uss(*arguments*) calculates the uncorrected sum of squares of the *arguments*.

var(*arguments*) calculates the variance of the *arguments*.

A.2.7   Random Number Functions

You can generate random numbers for various distributions using the following random number functions. The value of *seed* is any integer $\leq 2^{31} - 1$. If $seed \leq 0$, then the time of day is used to initialize the seed stream.

ranbin(*seed*, n, p) returns a Binomial(n, p) random variate.

rancau(*seed*) returns a Cauchy random variate.

ranexp(*seed*) returns an Exponential(1) variate.

rangam(*seed*, α) returns a Gamma(α, 1) variate.

rannor(*seed*) returns a $N(0, 1)$ variate.

ranpoi(*seed*, λ), where $\lambda > 0$, returns a Poisson(λ) random variate.

rantbl(*seed*, $p_1, \ldots, p_n$), where $0 \leq p_i \leq 1$ for $1 \leq i \leq n$, returns an observation generated from the probability mass function defined by p_1 through p_n.

rantri(*seed*, h), where $0 < h < 1$ returns an observation generated from the triangular distribution.

ranuni(*seed*) returns a number generated from the uniform distribution on the interval $(0, 1)$.

Appendix B

Arrays in the Data Step

Arrays are used in SAS when there is a group of variables we want to process in an identical way. For example, we may want to change 0 to . (missing) for every numeric variable in a data set. This is called *recoding*. Arrays are not permanent elements of a SAS data set. They exist only for the duration of the data step in which they are created. However, any changes made to elements of arrays are permanent for the variables they correspond to. Hence, arrays are treated somewhat differently in SAS than in other programming languages.

Arrays are created in **array** statements. For example, suppose a SAS data set contains the numeric variables x, y, and z and the character variables a and b. Then when dealing with this data set in a data step, the statements

```
array arr1{3} x y z;
array arr2{2} $ a b;
```

create two arrays: **arr1** is a numeric array of length 3 containing the variables x, y, and z, and **arr2** is a character array of length 2 containing the variables a and b.

The general form of the **array** statement is:

array *array-name*{*number of elements*} *list-of-variables*;

where *array-name* can be any valid SAS name. Instead of specifying the *number of elements*, we can use a * instead and then SAS determines this value from the list-of-variables. After an array has been formed, you can determine the number of elements in it using the **dim** function. For example,

```
d=dim(arr1);
```

213

```
put d;
```

prints 3 on the Log window. Note that we have to evaluate the function **dim** first.

Often, we want to refer to specific variables in an array. This is done using an *index*, as in *array-name*{*index-value*}. For example, arr1{1} refers to variable x and arr2{2} refers to variable b, so the statement

```
put arr1{1} arr2{2};
```

writes the values of x and b in the Log window. References to arrays can occur only after an array has been defined in the data step.

In our definition of an array the *index* runs from 1 up to **dim**(*array-name*). Sometimes it is more convenient to let the *index* run between other integer values. For example, if our data set contains the numeric variables x50, x51, , x61 and we want to define an array x containing these variables, then we could use

```
array x{12} x50-x61;
```

and x{i} refers to variable xj with $j = 49 + i$. More conveniently, we can specify a lower and upper bound for the index in the definition, as in

```
array x{50:61} x50-x61;
```

and now the index ranges between 50 and 61. To determine the lower and upper bounds of the index of an array, use the **lbound** and **hbound** functions. For example,

```
lb=lbound(x);
hb=hbound(x);
put lb hb;
```

writes 50 and 61 in the Log window when we define x with these lower and upper bounds. We have to evaluate the functions **lbound** and **hbound** first before we can output their values. Note that **dim**(*array-name*) = **hbound**(*array-name*) − **lbound**(*array-name*).

It is also possible to initialize arrays to take specific values via a statement of the form

array *array-name* {*number of elements*} *variable.list* (*list of initial values separated by blanks*);

For example,

```
array x{3} x1-x3 (0 1 2);
```

assigns the values x1 = 0, x2 = 1, and x3 = 2, and these variables are permanent elements of the data set (recall that arrays exist only for the duration of the particular data step in which they are defined). The variables x1-x3 are retained in the SAS data set, so they should be dropped if they are not needed.

Arrays are commonly used in do groups, allowing us to take advantage of the indexing feature to process many variables at once. For example, suppose we have 20 variables whose values are stored in a text file C:\datafile.txt with spaces between the values. Further, suppose these variables take values in {1, 2, ..., 10} but that 10 has been input into C:\datafile.txt as A. We want to treat the variables as numeric variables, but we don't want to edit C:\datafile.txt to change every A to a 10. The following program does this recoding.

```
data example (drop = t1-t10 x1-x20);
array test* $ t1-t10 ('1' '2' '3' '4' '5' '6' '7' '8'
    '9' 'A');
array x{*} $ x1-x20;
array y{*} y1-y20;
infile 'datafile';
input $ x1-x20;
do i=lbound(x) to hbound(x);
do j=1 to 10;
if (x{i} = test{j}) then y{i} = j;
end;
end;
```

Notice that we have dropped the character variables t1-t10 and x1-x20 from the data set **example**. This assumes that these variables will not be needed in any part of the SAS program that will make use of the data set **example**, and it is done to cut down the size of the data set. The program keeps the numeric variables y1-y20 in the data set. These take values in {1, 2, ..., 10}, and they are available for analysis by other SAS procedures. Further, notice the use of lbound(x) and hbound(x). We could have substituted 1 and 20, respectively, but the application of these functions in this program demonstrates a common usage.

The **array** statement has many other features. We discuss *explicit arrays*. *Implicit arrays*, used in older versions of SAS, are to be avoided now. Further, there are *multidimensional arrays*. For a discussion of these features, see reference 1 in Appendix E.

Appendix C

PROC IML

The procedure **proc iml** (Interactive Matrix Language) can be used for carrying out many of the calculations arising in linear algebra. Interactive Matrix Language (IML) is really a programming language itself. To invoke it we use the statement

```
proc iml;
```

within a SAS program. Various IML commands create matrices and perform operations on them. We describe here some of the main features of the IML language; the reader is referred to reference 7 in Appendix E for more details.

In IML matrices have the following characteristics:

- Matrices are referred to by SAS names; they must be from 1 to 8 characters in length, begin with a letter or an underscore, and consist only of letters, numbers, and underscores.

- They may contain elements that have missing values.

- The dimension of the matrix is defined by the number of rows and columns. An $m \times n$ matrix has mn elements — m rows and n columns.

C.1 Creating Matrices

C.1.1 Specifying Matrices Directly in IML Programs

A matrix can be specified in IML in a number of ways. The most direct method is to specify each element of the matrix. The dimensions of the matrix are determined by the way you punctuate the values. If there are multiple elements, i.e., the matrix is not 1×1, use braces { } to enclose the values, with elements in the same row separated by spaces, and rows separated by commas. If you use commas to create multiple rows, all rows must have the same number of elements. A period represents a missing numeric value. For example, the commands

```
proc iml;
c={1 2,3 4};
print c;
run;
```

creates the 2×2 matrix

$$c = \begin{pmatrix} 1 & 2 \\ 3 & 4 \end{pmatrix}$$

and prints it in the Output window. Scalars are matrices that have only one element. You define a scalar with the matrix name on the left-hand side and its value on the right-hand side. You do not need to use braces when there is only one element. A repetition factor can be placed in brackets before an element. For example, the statement

```
d = {[2] 1, [2] 2};
```

creates the matrix

$$d = \begin{pmatrix} 1 & 1 \\ 2 & 2 \end{pmatrix}$$

If you use the same name for two matrices, the latest assignment in the program overwrites the previous assignment.

C.1.2   Creating Matrices from SAS Data Sets

Sometimes you want to use data in a SAS data set to construct a matrix. Before you can access the SAS data set from within IML, you must first submit a **use** command to open it and make it the current data set for IML. The general form of the **use** statement is

use *SASdataset;*

where *SASdataset* names some data set. This is the current data set until a **close** command is issued to close the SAS data set. The form of the **close** statement is

close *SASdataset*;

where *SASdataset* names the data set.

Transferring data from a SAS data set, after it has been opened, to a matrix is done using the **read** statement. The general form of the **read** statement is

read <*range*><**var** *operand*> <**where**(*expression*)> <**into** *name*>;

where *range* specifies a range of observations, *operand* selects a set of variables, *expression* is a logical expression that is true or false, and *name* names a target matrix for the data.

The **read** statement with the **var** clause is used to read variables from the current SAS data set into column vectors of the **var** clause. Each variable in the **var** clause becomes a column vector. For example, the following program

```
data one;
input x y z;
cards;
1 2 3
4 5 6
proc iml;
use one;
read all var {x y} into a;
```

creates a data set called **one** containing two observations and three variables, x, y, and z. Then IML creates the 2×2 matrix

$$a = \begin{pmatrix} 1 & 2 \\ 4 & 5 \end{pmatrix}$$

from all the observations in this data set using variables x and y.

The following commands create the matrix

$$b = \begin{pmatrix} 1 & 2 \end{pmatrix}$$

by selecting only the observations where z = 3.

```
proc iml;
use one;
read all var {x y} into b where (z = 3);
```

If you want all variables to be used, use var_all_.

C.1.3   Creating Matrices from Text Files

Use the **infile** and **input** statements to read the contents of a text file into a SAS data set. Then invoke **proc iml** and use this data set to create the matrix as in section C.1.2. For example, if the file c:\stuff.txt has contents

```
1 2 3
4 5 6
```

then the statements

```
data one;
infile 'c:\stuff';
input x1 - x3;
proc iml;
use one;
read all var {x1 x2} into a;
```

create the matrix

$$a = \begin{pmatrix} 1 & 2 \\ 4 & 5 \end{pmatrix}$$

from the data in the file. The **closefile** statement, with the file pathname specified, closes files opened by an **infile** statement.

C.2   Outputting Matrices

C.2.1   Printing

As we have seen, the **print** statement writes a matrix in the Output window. For example, if a matrix C has been created, then

```
print c;
```

writes c in the Output window. It is possible to print a matrix with specified row and column headings. For example, if you have a 3×5 matrix, consisting of three students' marks for tests 1 through 5, then the statements

```
student = {student1 student2 student3};
test = {test1 test2 test3};
print results[rowname = student colname = test];
```

prints out the matrix named **results** using the row names given in the **student** vector and the column names given in the **test** vector. Note that the **print** statement can also print a message in the Output window by enclosing such a message in double quotes. Multiple matrices can be printed with the same **print** statement by listing them with a space between them.

C.2.2 Creating SAS Data Sets from Matrices

A SAS data set can be created from a matrix using the **create** and **append** statements. The columns of the matrix become the data set variables and the rows of the matrix become the observations. For example, an $n \times m$ matrix creates a SAS data set with m variables and n observations. Suppose we have created an $n \times 5$ matrix a in IML. The following statements

```
varnames = {x1 x2 x3 x4 x5};
create one from a [colname = varnames];
append from a;
```

create a SAS data set called **one** that has as its observations the n rows of matrix a and variables x1, x2, x3, x4, x5, which label the columns. Any other variable names could be used.

C.2.3 Writing Matrices to Text Files

Use the **file** and **put** commands to write a matrix into a text file. For example, the program

```
proc iml;
file 'c:\stuff';
do i = 1 to nrow(a);
do j = 1 to ncol(a);
y = a[i,j];
put y @;
```

```
    end;
    put;
    end;
```

writes the matrix a into the file `c:\stuff` row by row. The **closefile** statement, with the file pathname specified, closes files opened by a **file** statement.

C.3   Matrix Operations

Matrices can be created from other matrices by performing operations on matrices. These operations can be functions applied to single matrices, such as, "take the square root of every element in the matrix" or "form the inverse of a matrix," or they can operate on several matrices to produce a new matrix. Mathematical functions of a single real argument (see Appendix A for a listing) operate on a matrix elementwise. Matrix functions operate on the entire matrix. For example,

```
    a = sqrt(b);
```

assigns the square root of each element of matrix b to the corresponding element of matrix a, while the command

```
    x = inv(y);
```

calls the **inv** function to compute the inverse of the y matrix and assign the results to x.

C.3.1   Operators

There are three types of operators used in matrix expressions, or expressions involving one or more matrices.

prefix operators are placed in front of operands; e.g., -a reverses the sign of each element of a.

infix operators are placed between operands; e.g., a + b adds corresponding elements of the two matrices.

postfix operators are placed after an operand; e.g., a' uses the transpose operator ' to transpose a.

We define various matrix operators but first note here the order of precedence among the operators. Subscripts, ', -(prefix), ##, and ** have the highest priority, followed by *, #, <>, ><, /, and @, followed by + and −, followed by || and //, and finally <, <=, >, >=, = and ˆ=. We recommend using parentheses to prevent ambiguity. For example, `a*(b+c)` forms the matrix b + c and then premultiplies this matrix by a. When a missing value occurs in an operand, IML assigns a missing value to the result.

Addition +

The statements:

```
a = {1 2, 3 4};
b = {1 1, 1 1};
c = a + b;
```

produce the matrix

$$c = \begin{pmatrix} 2 & 3 \\ 4 & 5 \end{pmatrix}$$

You can also use the addition operator to add a matrix and a scalar or two scalars. When you add a matrix and a scalar, the scalar value is added to each element of the matrix to produce a new matrix. (When a missing value occurs in an operand, IML assigns a missing value for the corresponding element in the result.)

Comparison

The comparison operators include <, >, <=, >=, ˆ= (not equal to). The comparison operators compare two matrices element by element and produce a new matrix that contains only zeros and ones. If an element comparison is true, the corresponding element of the new matrix is 1. Otherwise, the corresponding element is 0.

Concatenation ||

This operator produces a new matrix by horizontally joining two matrices. The statements

```
a = {1 2, 4 5};
b = {0, 8};
c = a || b;
```

produce the matrix

$$c = \begin{pmatrix} 1 & 2 & 0 \\ 4 & 5 & 8 \end{pmatrix}$$

Concatenation, Vertical //

This operator produces a new matrix by vertically joining two matrices. The format is the same as that of the horizontal operator.

Direct (Kronecker) Product @

The direct product operator @ produces a new matrix that is the direct product of two matrices. If we have two matrices called `matrix1` and `matrix2`, then the statement

```
matrix3 = matrix1 @ matrix2;
```

takes the direct product of the two matrices and calls it `matrix3`.

Division /

The division operator in the statement

```
c = a / b;
```

divides each element of the matrix `a` by the corresponding element of the matrix `b` producing a matrix of quotients. You can also use this operator to divide a matrix by a scalar. If either operand is a scalar, the operation does the division for each element and the scalar value. Once again, when a missing value occurs in an operand, IML assigns a missing value for the corresponding element in the result.

Element Maximum <>

The operator <> in the statement

```
c = a <> b;
```

compares each element of a to the corresponding element of b. The larger of the two values becomes the corresponding element of the new matrix. When either argument is a scalar, the comparison is between each matrix element and the scalar.

Element Minimum ><

This operator works similarly to the element maximum operator, but it selects the smaller of two elements.

Multiplication, Elementwise

This operator produces a new matrix whose elements are the products of the corresponding elements of two matrices. For example, the statements

```
a = {1 2, 3 4};
b = {9 8, 7 6};
c = a # b;
```

produce the matrix

$$c = \begin{pmatrix} 9 & 16 \\ 21 & 24 \end{pmatrix}$$

Multiplication, Matrix *

This operator performs matrix multiplication. The first matrix must have the same number of columns as the second matrix has rows. The new matrix has the same number of rows as the first matrix and the same number of columns as the second matrix. For example, the statements

```
a = {1 2, 3 4};
b = {1 2};
c = b * a;
```

produce the matrix

$$c = \begin{pmatrix} 7 & 10 \end{pmatrix}$$

Power, elementwise

This operator creates a new matrix whose elements are the elements of the first matrix specified raised to the power of the corresponding elements of the second matrix specified. If one of the operands is a scalar, then the operation takes the power for each element and the scalar value. The statements

```
a = {1 2 3};
b=a##3;
```

produce the matrix

$$b = \begin{pmatrix} 1 & 8 & 27 \end{pmatrix}$$

Power, matrix **

This operator raises a matrix to a power. The matrix must be square, and the scalar power that the matrix is raised to must be an integer greater than or equal to -1. The statements

```
a = {1 2, 1 1};
b = a ** 2;
```

produce the matrix

$$b = \begin{pmatrix} 3 & 4 \\ 2 & 3 \end{pmatrix}$$

Note that this operator can produce the inverse of a matrix; `a ** (-1)` is equivalent to `inv(a)`.

Sign Reverse −

This operator produces a new matrix whose elements are formed by reversing the sign of each element in a matrix. A missing value is assigned if the element is missing. The statements

```
a = {-1 2, 3 4};
b = -a;
```

produce the matrix

$$b = \begin{pmatrix} 1 & -2 \\ -3 & -4 \end{pmatrix}$$

Subscripts [ ]

Subscripts are postfix operators, placed in square brackets [ ] after a matrix operand, that select submatrices. They can be used to refer to a single element of a matrix, refer to an entire row or column of a matrix, or refer to any submatrix contained within a matrix. If we have a 3×4 matrix x, then the statements

```
x21 = x[2, 1];
print x21;
```

print the element of x in the second row and first column. The statements

```
firstrow = x[1,];
print firstrow;
```

print the first row of x, and the statements

```
firstcol = x[,1];
print firstcol;
```

print the first column of x. You refer to a submatrix by the specific rows and columns you want. Include within the brackets the rows you want, a comma, and the columns you want. For example, if y is a 4 × 6 matrix, then the statements

```
submat = y[{1 3},{2 3 5}];
print submat;
```

print out a matrix called submat, consisting of the first and third rows of y and the second, third, and fifth columns of y.

Subtraction -

This operator in the statement

```
c = a - b;
```

produces a new matrix whose elements are formed by subtracting the corresponding elements of the second specified matrix from those of the first specified matrix. You can also use the subtraction operator to subtract a matrix and a scalar. If either argument is a scalar, the operation is performed by using the scalar against each element of the matrix argument.

Transpose '

The transpose operator ' exchanges the rows and columns of a matrix. The statements

```
a = {1 2, 3 4, 5 6};
b = a';
```

produce the matrix

$$b = \begin{pmatrix} 1 & 3 & 5 \\ 2 & 4 & 6 \end{pmatrix}$$

C.3.2   Matrix-generating Functions

Matrix-generating functions are functions that generate useful matrices.

block

The **block** function creates a block-diagonal matrix from the argument matrices. For example if a and b are matrices then

    c=block(a,b);

is the matrix

$$c = \begin{pmatrix} A & 0 \\ 0 & B \end{pmatrix}$$

design

The **design** function creates a design matrix from a column vector. Each unique value of the vector generates a column of the design matrix. This column contains ones in elements whose corresponding elements in the vector are the current value; it contains zeros elsewhere. For example, the statements

    a = {1, 1, 2, 3};
    a = design (a);

produce the design matrix

$$a = \begin{pmatrix} 1 & 0 & 0 \\ 1 & 0 & 0 \\ 0 & 1 & 0 \\ 0 & 0 & 1 \end{pmatrix}$$

i

The **i** function creates an identity matrix of a given size. For example,

    Ident = i(4);

creates a 4×4 identity matrix named `Ident`.

Index operator :

Using the index operator : creates *index vectors*. For example, the statement

    r = 1:5;

produces an index vector r

$$r = (\ 1\quad 2\quad 3\quad 4\quad 5\ )$$

j

The **j** function creates a matrix of a given dimension. This function has the general form **j**(*nrow, ncol, value*), and it creates a matrix having *nrow* rows, *ncol* columns and all element values are equal to *value*. The *ncol* and *value* arguments are optional, but you will usually want to specify them. The statement

    a = j(1, 5, 1);

creates a 1×5 row vector of 1's.

C.3.3 Matrix Inquiry Functions

Matrix inquiry functions return information about a matrix.

all

The **all** function is used when a condition is to be evaluated in a matrix expression. The resulting matrix is a matrix of 0's, 1's, and possibly missing values. If all values of the result matrix are nonzero and nonmissing, the condition is true; if any element in the resulting matrix is 0, the condition is false.

any

The **any** function returns a value of 1 if any of the elements of the argument matrix are nonzero and a value of 0 if all of the elements of the matrix are zeros.

loc

The **loc** function creates a $1 \times n$ row vector, where n is the number of nonzero elements in the argument. Missing values are treated as zeros. The values in the resulting row vector are the locations of the nonzero elements in the argument. For example,

```
a = {1 2 0, 0 3 4};
b = loc(a);
```

produce the row vector

$$b = \begin{pmatrix} 1 & 2 & 5 & 6 \end{pmatrix}$$

ncol

The **ncol** function provides the number of columns of a given matrix.

nrow

The **nrow** function provides the number of rows of a given matrix.

C.3.4   Summary Functions

Summary functions return summary statistics on the matrix.

max

The **max** function produces a single numeric value that is the largest element in all arguments. There can be as many as 15 argument matrices.

min

The **min** function returns a scalar that is the smallest element in all arguments.

ssq

The **ssq** function returns a scalar containing the sum of squares for all the elements of all arguments. The statements

```
a = {1 2 3, 4 5 6};
```

```
x = ssq(a);
```

result in x having a value of 91.

sum

The **sum** function returns a scalar that is the sum of all elements in all arguments.

C.3.5 Matrix Arithmetic Functions

Matrix arithmetic functions perform matrix algebraic operations on the matrix.

cusum

The **cusum** function returns a matrix of the same dimension as the argument matrix. The result contains the cumulative sums obtained by scanning the argument and summing in row-major order. For example,

```
b = cusum({5 6, 3 4});
```

produces the matrix

$$b = \begin{pmatrix} 5 & 11 \\ 14 & 18 \end{pmatrix}$$

hdir

The **hdir** function has the general form **hdir**(*matrix1*, *matrix2*). This function performs a direct product on all rows of *matrix1* and *matrix2* and creates a new matrix by stacking these row vectors into a matrix. The arguments, *matrix1* and *matrix2* must contain the same number of rows. The resulting matrix has the same number of rows as *matrix1* and *matrix2*, and the number of columns is equal to the product of the number of columns in *matrix1* and *matrix2*.

trace

The **trace** function returns the trace of the matrix. That is, it sums the diagonal elements of a matrix. The statement

```
a = trace({5 2, 1 3});
```

produces the result

$$a = 8$$

C.3.6   Matrix Reshaping Functions

Matrix reshaping functions manipulate the matrix and produce a reshaped matrix.

diag

The **diag** function returns a diagonal matrix whose diagonal elements are the same as those of the matrix argument, and all elements not on the diagonal are zeroes.

insert

The **insert** function inserts one matrix inside another matrix. This function takes the form **insert**$(x, y,$ *row, column*$)$, where x is the target matrix, y is the matrix to be inserted into the target matrix, row is the row where the insertion is to be made, and column is the column where the insertion is to be made. For example, the statements

```
a = {1 2, 3 4};
b = {5 6, 7 8};
c = insert(a, b, 2, 0);
```

produce the matrix

$$c = \begin{pmatrix} 1 & 2 \\ 5 & 6 \\ 7 & 8 \\ 3 & 4 \end{pmatrix}$$

rowcat

The **rowcat** function returns a one-column matrix with all row elements concatenated into a single string. This function has the form **rowcat**(*matrix, rows, columns*) where *rows* select the rows of the matrix and *columns* select the columns of the matrix.

t

The **t** function returns the transpose of the argument matrix. It is equivalent to the transpose postfix operator '. This function has the form **t**(*matrix*).

vecdiag

The **vecdiag** function returns a column vector containing the diagonal elements of the argument matrix. The matrix must be square to use this function.

C.3.7 Linear Algebra and Statistical Functions

Linear algebra and statistical functions perform algebraic functions on the matrix.

covlag

The **covlag** function computes a sequence of lagged cross-product matrices. This function is useful for computing sample autocovariance sequences for scalar or vector time series.

det

The **det** function computes the determinant of a square matrix. This function has the form **det**(*matrix*).

echelon

The **echelon** function uses elementary row operations to reduce a matrix to row-echelon normal form.

eigval

The **eigval** function returns a column vector of the eigenvalues of a matrix. The eigenvalues are arranged in descending order. The function has the form **eigval**(*matrix*), where *matrix* is a symmetric matrix.

eigvec

The **eigvec** function returns a matrix containing eigenvectors of a matrix. The columns of the resulting matrix are the eigenvectors. The function has the form **eigvec**(*matrix*), where *matrix* is a symmetric matrix.

fft

The **fft** function performs the finite Fourier transform.

ginv

The **ginv** function returns the matrix that is the generalized inverse of the argument matrix. This function has the form **ginv**(*matrix*), where *matrix* is a numeric matrix or literal.

hankel

The **hankel** function generates a Hankel matrix. This function has the form **hankel**(*matrix*), where *matrix* is a numeric matrix.

hermite

The **hermite** function uses elementary row operations to reduce a matrix to Hermite normal form. For square matrices, this normal form is upper-triangular and idempotent. If the argument is square and nonsingular, the result is the identity matrix. This function has the form **hermite**(*matrix*).

homogen

The **homogen** function solves the homogeneous system of linear equations $AX = 0$ for X. This function has the form **homogen**(*matrix*), where *matrix* is a numeric matrix.

ifft

The **ifft** function computes the inverse finite Fourier transform.

inv

The **inv** function produces a matrix that is the inverse of the argument matrix. This function has the form **inv**(*matrix*), where *matrix* is a square, nonsingular matrix.

polyroot

The **polyroot** function finds zeros of a real polynomial and returns a matrix of roots. This function has the form **polyroot**(*vector*), where *vector* is an $n \times 1$ (or $1 \times n$) vector containing the coefficients of an $(n-1)$ degree polynomial with the coefficients arranged in order of decreasing powers. The **polyroot** function returns an $(n-1) \times 2$ matrix containing the roots of the polynomial. The first column of the matrix contains the real part of the complex roots and the second column contains the imaginary part. If a root is real, the imaginary part will be 0.

rank

The **rank** function ranks the elements of a matrix and returns a matrix whose elements are the ranks of the corresponding elements of the argument matrix. The ranks of tied values are assigned arbitrarily rather than averaged. For example, the statements

```
a = {2 2 1 0 5};
b = rank(a);
```

produce the vector

$$b = \begin{pmatrix} 3 & 4 & 2 & 1 & 5 \end{pmatrix}$$

ranktie

The **ranktie** function ranks the elements of a matrix using tie-averaging, and returns a matrix whose elements are the ranks of the corresponding elements of the argument matrix. For example, the statements

```
a = {2 2 1 0 5};
b = ranktie(a);
```

produce the vector

$$b = \left(\begin{array}{ccccc} 3.5 & 3.5 & 2 & 1 & 5 \end{array} \right)$$

root

The **root** function performs the Cholesky decomposition of a matrix and returns an upper triangular matrix. This function has the form **root**(*matrix*), where *matrix* is a positive-definite matrix.

solve

The **solve** function solves the set of linear equations $AX = B$ for X. This function has the form **solve**(*A,B*), where A is an $n \times n$ nonsingular matrix and B is an $n \times p$ matrix. The statement x = solve(a,b); is equivalent to the statement x = inv(a)*b; but the **solve** function is recommended over the **inv** function because it is more efficient and accurate.

sweep

The **sweep** function has the form **sweep**(*matrix, index-vector*). The **sweep** function sweeps *matrix* on the pivots indicated in *index-vector* to produce a new matrix. The values of the *index-vector* must be less than or equal to the number of rows or the number of columns in *matrix*, whichever is smaller.

toeplitz

The **toeplitz** function generates a Toeplitz matrix. This function has the form **toeplitz**(*A*), where A is a matrix.

C.4   Call Statements

Call statements invoke a subroutine to perform calculations, or operations. They are often used in place of functions when the operation returns multiple results. The general form of the **call** statement is

call *subroutine* (*arguments*);

where *arguments* can be matrix names, matrix literals, or expressions. If you specify several arguments, use commas to separate them.

armacov

The **armacov** call statement computes an autocovariance sequence for an ARMA model.

armalik

The **armalik** call statement computes the log likelihood and residuals for an ARMA model.

eigen

The **eigen** call statement has the form **call eigen**(*eigenvalues, eigenvectors, symmetric-matrix*);, where *eigenvalues* names a matrix to contain the eigenvalues of the input matrix, *eigenvectors* names a matrix to contain the eigenvectors of the input matrix, and *symmetric-matrix* is a symmetric, numeric matrix. The **eigen** subroutine computes *eigenvalues*, a column vector containing the eigenvalues of the argument matrix, in descending order. It also computes *eigenvectors*, a matrix containing the orthonormal column eigenvectors of the argument matrix, arranged so that the first column of eigenvectors is the eigenvector corresponding to the largest eigenvalue, and so on.

geneig

The **geneig** call statement computes eigenvalues and eigenvectors of the generalized eigenproblem. It has the form **call geneig**(*eigenvalues, eigenvectors, symmetric-matrix1, symmetric-matrix2*);, where *eigenvalues* is a returned vector containing the eigenvalues, *eigenvectors* is a returned matrix containing the corresponding eigenvectors, *symmetric-matrix1* is a symmetric numeric matrix, and *symmetric-matrix2* is a positive definite matrix. The statement

```
call geneig (m, e, a, b);
```

computes eigenvalues M and eigenvectors E of the generalized eigenproblem

$$AE = BE \operatorname{diag}(M)$$

gsorth

The **gsorth** call statement computes the Gram-Schmidt factorization. It has the form **call gsorth**$(Q, R, lindep, A)$, where A is an $m \times n$ input matrix with $m \geq n$, Q is an $m \times n$ column orthonormal output matrix, R is an upper triangular $n \times n$ output matrix, and *lindep* is an output flag with a value of 0 if columns of A are independent and a value of 1 if they are dependent. The **gsorth** subroutine computes the column-orthonormal $m \times n$ matrix Q and the upper-triangular $n \times n$ matrix R such that

$$A = QR$$

ipf

The **ipf** call statement performs an iterative proportional fit of the marginal totals of a contingency table.

lp

The **lp** call statement solves the linear programming problem.

svd

The **svd** call statement computes the singular value decomposition. It has the form **call svd**$(U, Q, V, A);$, where U, Q and V are the returned decomposition matrices, and A is the input matrix that is decomposed. The **svd** subroutine decomposes a real $m \times n$ matrix A(where m is greater than or equal to n) into the form

$$A = U \operatorname{diag}(Q) V`.$$

C.5   Control Statements

IML is a programming language. It has many features that allow you to control the path of execution through the statements.

abort and stop

The **abort** statement stops execution and exits from IML. If there are additional procedures in the SAS program, they will be executed. The **stop** statement stops execution but it does not cause an exit from IML.

if-then-else

The general form of the **if-then-else** statement is

if *expression* **then** *statement1*;
else *statement2*;

where *expression* is a logical expression and *statement1 and statement2* are executable IML statements. The truth value of the **if** *expression* is evaluated first. If *expression* is true then *statement1* is executed and *statement2* is ignored. If *expression* is false, then *statement2* in the **else** statement, if **else** is present, is executed. Otherwise, the next statement after the **if-then** statement is executed. For example,

```
if max(a)<20 then p=0;
else p=1;
```

determines whether the largest value in the matrix **a** is less than 20. If it is, then **p** is set to 0. Otherwise, **p** is set to 1.

do

The **do** statement specifies that the statements following the **do** statement are executed as a group until a matching **end** statement appears. These statements group statements as a unit. The **do** statement is always accompanied by a corresponding **end** statement and would appear in a program as

do;
statements
end;

where *statements* consists of a set of IML statements.

Often **do** statements appear in **if-then-else** statements, where they designate groups of statements to be performed when the **if** condition is true or false. For example, in

```
if x=y then;
do;
i=i+1;
print x;
end;
print y;
```

the statements between the **do** and the **end** statements are performed only if x=y. If this is not the case, then the statements in the **do** group are skipped and the next statement is executed.

Statements between the **do** and the **end** statements can be executed repetitively when the **do** group has the form

do *variable* = *start* **to** *stop* <**by** *increment*>;

where *variable* is the name of a variable indexing the loop, *start* is the starting value for the looping variable, *stop* is the stopping value for the looping variable, and *increment* is an increment value. The *start, stop,* and *increment* values should be scalars. The variable is given a new value at the end of each repetition of the group. It starts with the *start* value and is incremented by the *increment* value after each iteration. The iterations continue as long as the variable is less than or equal to the *stop* value. The statements

```
do i=1 to 5 by 2;
print i;
end;
```

produce the output

```
1
3
5
```

The **until** expression makes the conditional execution of a set of statements possible. This occurs in one of the forms

do until(*expression*);
do *variable* = *start* **to** *stop* <**by** *increment*> **until**(*expression*);

where *expression* is a logical expression whose truth value is evaluated at the bottom of the loop, *variable* is the name of the variable indexing the loop, *start* is the starting value for the looping variable, and *increment* is the increment value for *variable*. In the statements

```
x=1;
do until (x > 100);
x+1;
end;
print x;
```

the body of the loop executes until the value of x exceeds 100 so that the value x=101 is printed.

Using a **while** expression also makes possible the conditional execution of a set of statements iteratively. This occurs in one of the forms

do while(*expression*);
do *variable = start* **to** *stop* <**by** *increment*> **while**(*expression*);

where *expression* is a logical expression whose truth value is evaluated at the top of the loop and *variable, start, stop,* and *increment* are as previously described. The **while** *expression* is evaluated at the top of the loop, and the statements inside the loop are executed repeatedly as long as *expression* yields a nonzero or nonmissing value. The statements

```
x=1;
do while (x<100);
x=x + 1;
end;
print x;
```

start with x=1 and add 1 to x until x=100 and then this value is printed.

goto and link

During normal execution, statements are executed one after another. The **goto** and **link** statements instruct IML to jump from one part of a program to another. The place to which execution jumps is identified by a label, which is a name followed by a colon placed before an executable statement. You can program a jump by using either the **goto** statement or the **link** statement. These take the form

goto *label*;
link *label*;

where *label* appears elsewhere in the program followed by a colon. Both the **goto** and **link** statements instruct IML to jump immediately to the labeled statement. However, the **link** statement reminds IML where it jumped from so that execution can be returned there if a **return** statement is encountered. The **link** statement provides a way of calling sections of code as if they were subroutines. The **link** statement calls the routine. The routine begins with the *label* and ends with a **return** statement. Note that the **goto** and **link** statements must be located within a **do** group or module (see the following discussion). Below are examples of the **goto** and **link** statements within **do** groups. The statements

```
do;
if x <0 then goto negative;
y=sqrt(x);
print y;
stop;
negative:
print ''Sorry,X is negative'';
end;
```

cause an error message to be printed if we try to take the square root of a negative number. And the statements

```
do;
if x < 0 then link negative;
y = sqrt(x);
print y;
stop;
negative:
print ''Using Abs.  value of negative X'';
x = abs(x);
return;
end;
```

compute the square root of a nonnegative number and the absolute value of any negative number.

C.6   Modules

Modules are used for creating a group of statements that can be invoked as a unit from anywhere in the program, that is, these statements constitute a

subroutine or function. A module always begins with the **start** statement
and ends with the **finish** statement. If a module returns a single parameter,
then it is called a function and it is executed as if it were a built-in IML
function. Otherwise, a module is called a subroutine, and you execute the
module with either the **run** statement or the **call** statement. Of course,
modules must appear in the program before they are used.

To write a function module, include a **return** statement that assigns the
returned value to a variable. This statement is necessary for a module to be
a function. The following statements

```
start add(x,y);
sum = x+y;
return(sum);
finish;
```

define a function called **add** for adding two arguments and assigns the returned value to a variable called **sum**.

The statements

```
start mymod(a, b, c);
a = sqrt (c);
b = log (c);
finish;
```

define a module named **mymod** that returns matrices containing the square
root and log of each element of the argument matrix c. The subsequent
statement

```
run mymod (s, l, x);
```

causes the module **mymod** to be executed. Execution of the module statements
creates matrices s and l, containing the square roots and logs, respectively,
of the elements of x.

The **pause** statement stops execution of the module and remembers
where it stopped executing. It can be used to prompt a user for input.
A **resume** statement allows you to continue execution at the place where
the most recent **pause** statement was executed.

C.7   Simulations Within IML

All the random number functions are available in IML, so we can write programs in IML to do simulations. This is convenient when we want to run a simulation where a substantial amount of the computation involves linear algebra. For example, the program

```
proc iml;
seed = 1267824;
Z = J(10, 3, 1);
do i = 1 to 10;
do j = 1 to 3;
Z[i,j] = rannor(seed);
end;
end;
```

generates a 3×10 matrix Z containing independent $N(0, 1)$ values. Note that we must first create a matrix Z before we can assign a value to any element of it. This is the role of the Z = J(3,10,1) statement. This statement must come before any other do loops if we want to repeatedly generate such matrices.

Appendix D

Advanced Statistical Methods in SAS

This manual covers many of the features available in SAS but there are a number of other useful techniques, while not relevant to an elementary course, are extremely useful in a variety of contexts. The material in this manual is good preparation for using the more advanced techniques.

We list here some of the more advanced methods available in SAS and refer the reader to references 3 and 4 in Appendix E for details.

General linear model and general ANOVA
Variance components and mixed models
MANOVA
Principal components
Factor analysis
Discriminant analysis
Cluster analysis
Nonlinear least-squares analysis
Time series analysis
Quality control tools
Survival analysis
Design of experiments

Appendix E

References

This manual is based on the following manuals published by SAS Institute Inc., SAS Circle, Box 8000, Cary, NC 27512-8000.

1. SAS Language Reference, Version 6, First Edition.

2. SAS Procedures Guide, Version 6, Third Edition.

3. SAS/STAT User's Guide Volume 1, Version 6, Fourth Edition.

4. SAS/STAT User's Guide Volume 2, Version 6, Fourth Edition.

5. SAS/GRAPH Software, Volume 1, Version 6, First Edition.

6. SAS/GRAPH Software, Volume 2, Version 6, First Edition.

7. SAS/IML Software, Version 6, First Edition.

8. SAS/QC(R) Software: Usage and Reference, Version 6, First Edition, Volume 1.

9. SAS/QC(R) Software: Usage and Reference, Version 6, First Edition, Volume 2.

Index

abort, 239
all, 229
any, 229
append, 221
arithmetic expression, 17
arithmetic operators, 17, 207
armacov, 237
armalik, 237
array, 213
arrays, 213
ascending order, 38
assignment statement, 16
axis, 64, 67

balanced data, 173
bar charts, 60
binomial distribution, 114
block, 228
block charts, 60
Bonferonni t tests, 175
boxplot, 60
boxplots from **proc gplot**, 78
boxplots, side by side, 185
by, 38, 53
by group processing, 40

call statements, 236
cards, 10
character variables, 16
chi-square distribution, 131
chi-square test, 153

class, 53
class variables, 170
close, 219
closefile, 220
column output, 28
comments, 13
comparison operators, 17, 208
concatenation operator, 208
control charts, 115
correlation coefficient, 78
correlation matrix, 81
coverage probability, 128
covlag, 233
create, 221
cross tabulation, 88
cumulative distribution, 46
cusum, 231

data, 53
data step, 13
datalines, 10
delimiter, 23
density curve of the $N(\mu, \sigma)$, 67
descending, 39
descending order, 38
design, 228
det, 233
diag, 232
dim, 213
direct product, 224
display manager window, 5

distribution-free methods, 139, 189
do-until, 35
do while, in IML, 241
do-end, 34
do-end until, 35
do-end while, 35
do-end, in IML, 239
do-until, in IML, 240
do-while, 35
drop, 21
Duncan's multiple range test, 176

echelon, 233
eigen, 237
eigenvalues, 237
eigval, 233
eigvec, 234
empirical distribution function, 46,
49

F distribution, 143
family error rate, 175
fft, 234
file, 29
Fisher's LSD, 175
format, 28
formats, 28
formatted input, 24
formatted output, 28
freq, 59
freq, 53
frequency, 46

gamma function, 131
geneigen, 237
generalized eigenvalues, 237
ginv, 234
goto, 33
goto, in IML, 241

Gram-Schmidt factorization, 238
grouped informat lists, 25
grouping, 46
gsorth, 238

hankel, 234
hbar, 61
hbound, 214
hdir, 231
help, 7
hermite, 234
high-resolution plots, 66
histogram, 62
homogen, 234

i, 228
if-then-else, 32
if-then-else, in IML, 239
ifft, 234
IML, 217
index operator :, 229
individual error rate, 175
infile, 23
infix operators, 222
informat, 24
informat, 26
input, 18
insert, 232
interleaving data sets, 20
interpol, 78
inv, 235
ipf, 238
iterative do, 34
iterative proportional fit, 238

j, 229

keep, 21
Kronecker product, 224

Kruskal-Wallis test, 193

lbound, 214
libname, 22, 30
linear origramming, 238
link, in IML, 241
list, 28
list input, 23
list output, 28
loc, 230
log odds, 198
log window, 5
logical expression, 17
logical operators, 17, 208
logistic regression, 197
logit link function, 197
lostcard, 26
lp, 238

Mann-Whitney statistic, 191
master data set, 32
matched pairs design, 138
matrices in IML, 217
matrix addition, 223
matrix comparisons, 223
matrix division, 224
matrix elementwise multiplication, 225
matrix elementwise power, 225
matrix horizontal concatenation, 223
matrix maximum, 224
matrix minimum, 225
matrix multiplication, 225
matrix operators, 222
matrix power, 226
matrix subscripts, 226
matrix subtraction, 227
matrix transpose, 227

matrix vertical concatenation, 224
max, in IML, 230
maximum operator, 208
menu bar, 5
menu commands, 4
merge, 20
midpoints, 64
miltiple regression, 167
min, in IML, 230
minimum operator, 208
missing, 16
missing, 26
model, 84
modules, 242

ncol, 230
noncentral chi-square distribution, 179
noncentral F distribution, 179
noncentral Student distribution, 139
nonpaprametric analyses, 189
nonparametric, 139
nonparametric methods, 189
noprint, 49, 56
normal, 59
normal probability plot, 60
normal quantile plot, 69
notsorted, 40
nrow, 230
numeric variables, 16
nway, 55

odds, 198
one-to-one merging, 21
one-way ANOVA model, 173
operator priority, 208
operators, 17, 207
options, 51

output, 28
output, 29, 54
output window, 7

p-th percentile, 68
pad, 24
pchart, 118
Pearson correlation coefficient, 80
pie charts, 60
plot, 59
pointer, 23
polyroot, 235
population distribution, 100
postfix operators, 222
power, 128
predicted, 85
prefix operators, 222
print, in IML, 220
priority of operators, 17
probability functions, 210
probit, 69
probit link function, 198
probnorm, 68
proc, 9
proc anova, 170
proc chart, 60
proc contents, 31
proc corr, 79
proc freq, 46, 87
proc gchart, 67
proc glm, 170
proc gplot, 77
proc iml, 217
proc logistic, 197
proc means, 51
proc nparlway, 189
proc plan, 98
proc plot, 75

proc print, 37
proc reg, 83
proc shewhart, 115
proc sort, 38
proc tabulate, 91
proc timeplot, 65
proc ttest, 141
proc univariate, 56
program editor window, 5
put, 27

random number functions, 212
random permutations, 98
range list, 16
rank, 235
ranktie, 235
read, 219
recoding, 213
proc reg, 159
relative frequency, 46
rename, 21
repeated sampling, 100
residual, 85
retain, 106
return, 33
root, 236
rowcat, 232
run, 11

sampling from distributions, 100
sampling without replacement, 98
SAS constants, 15
SAS data set, 14
SAS expression, 17
SAS functions, 17, 209
SAS name, 14
SAS statements, 9
SAS system viewer, 13

SAS variables, 16
SAS/Graph, 66
SAS/QC, 115
scatterplot, 73
Scheffe's multiple comparison procedure, 176
select-otherwise, 36
sequential analysis of variance, 171
set, 19
sign test, 139
simple do, 34
simulation, 107
simulations for confidence intervals, 126
simulations, in IML, 244
singular value decomposition, 238
solve, 236
special functions, 210
spectral decomposition, 237
ssq, in IML, 230
standard error of the estimate, 108
standardize, 68
standardized residuals, 161
statistical functions, 211
stem-and-leaf plot, 60
stop, 33
stop, in IML, 239
storage, 22
Student distribution, 135
studentized residuals, 161
subsetting if, 20
sum, in IML, 231
svd, 238
sweep, 236
symbol, 77

t confidence interval, 136
t test, 137

proc freq, 153
tables, 48
task bar, 5
title, **51**
toeplitz, 236
trace, 231
transactions data set, 32
trignometric functions, 210
truncation functions, 209
Tukey's studentized range test, 176
two-sample t confidence interval, 141
two-sample t test, 141
two-sample z confidence interval, 140
two-sample z test, 140
two-way ANOVA model, 181
Type I Sums of Squares, 171
Type III of Sums of Squares, 171

update, 31
use, 218

value, 78
var, 53
vbar, 61
vecdiag, 233

weight, 54
weight, 49
weighted correlation coefficient, 81
Wilcoxon rank sum statistic, 191
Wilcoxon rank sum test, 190
Wilcoxon signed rank statistic, 192

xbar chart, 116

z confidence interval, 124
z test, 125